MASS SPECTROMETRY FOR BIOTECHNOLOGY

MASS SPECTROMETRY FOR BIOTECHNOLOGY

Gary Siuzdak

The Department of Chemistry
The Scripps Research Institute
La Jolla, California

ACADEMIC PRESS
San Diego New York Boston London
Sydney Tokyo Toronto

This book is printed on acid-free paper. ♾

Academic Press
A Harcourt Science and Technology Company
525 B Street, Suite 1900, San Diego, California 92101-4495, USA
http://www.academicpress.com

Academic Press
Harcourt Place, 32 Jamestown Road, London NW1 7BY, UK
http://www.academicpress.com

Library of Congress Cataloging-in-Publication Data

Siuzdak, Gary.
Mass spectrometry for biotechnology / by Gary Siuzdak.
p. cm.
Includes index.
ISBN 0-12-647471-0 (pbk.)
1. Mass spectrometry. 2. Proteins--Analysis. 3. Nucleic acids--Analysis. 4. Biotechnology. I. Title.
QP519.9.M3S58 1995
574.19'245'028--dc20 95-30593
CIP

PRINTED IN THE UNITED STATES OF AMERICA
00 01 QW 9 8 7 6 5

Contents

Preface

In the past, mass spectrometry was confined to the realm of small molecules; large molecules did not survive the desorption and ionization process intact. More recently, the development of new "mild" desorption and ionization methods has revolutionized the analysis of large biomolecules, making mass spectrometry an important analytical tool for biological research. With these new ionization technologies and the entry of mass spectrometry into the realm of biomolecules, the need for a descriptive book on this topic has become apparent.

Three new mass spectrometry technologies have made the study of biomolecules routine: matrix-assisted laser desorption/ionization (MALDI), electrospray ionization (ESI), and fast atom/ion bombardment (FAB). These technologies have allowed for the analysis of proteins, peptides, carbohydrates, oligonucleotides, and drug metabolites, with detection capabilities ranging from the picomole to the femtomole level. In addition, these techniques can provide molecular weight accuracy on the order of $\pm 0.01\%$ and structural information as well.

The challenge of becoming proficient in these techniques requires learning some of the basics of mass analysis and ionization methods and then identifying which techniques are most appropriate for specific problems. For example, a biochemist requiring mass information on large heterogeneous glycoproteins would probably be most successful with MALDI, while a protein chemist routinely analyzing homogeneous proteins in the mass range 20–70 kDa will find the highest mass accuracy with ESI. This book has therefore been written to familiarize scientists with the general principles of mass spectrometry and its utility as a research tool.

During the past decade mass spectrometry has seen increasing application in both the chemical and the biological sciences. In fact, the potential applications in biology have been a strong motivation, both intellectually and economically, behind recent advances in this technology. These advances have resulted in commercially available and increasingly affordable instruments with the ability to analyze bio-ions of mass greater than 200,000 Da, to study noncovalent interactions, to sequence peptides and even proteins through direct analysis of complex Edman degradation mixtures, and to monitor intracellular drug interactions directly. In general, these new developments signify another dimension in molecular characterization through a new level of sensitivity, accuracy, and mass range.

This book presents information about the operational principles, ionization methods, and analysis capabilities in a format targeted to biologists, molecular biologists, and biochemists. The basics of biological mass spectrometry have been described in order to facilitate understanding, with illustrations to facilitate presentation. The chapters include the descriptions of ionization sources (Chapter 1), mass analyzers and detectors (Chapter 2), general questions (Chapter 3), peptides and proteins (Chapter 4), other biomolecules (Chapter 5), and specific examples taken from *Science, Proceedings of the National Academy of Sciences, Nature, Analytical Biochemistry, Journal of the American Chemical Society,* and *Biochemistry* (Chapter 6). Also included are basic terminology, definitions, and practical tips on using matrix-assisted laser desorption, electrospray ionization, and fast atom/ion bombardment.

Acknowledgments

I gratefully acknowledge my friends and colleagues who helped review and edit this manuscript. I express my sincere appreciation to members of the Scripps Mass Spectrometry Laboratory, Jennifer Boydston, Thomas Knapp, Brian Bothner, Kelly S. Chatman, Alanna Phillips, Anna Aparicio, and Mark Feinstein, who spent many hours reviewing, discussing, editing, and conducting research throughout the course of this writing. Jennifer and Thomas have made especially worthwhile contributions to this manuscript.

The following individuals have provided extremely helpful critiques and suggestions:

Dr. Michael Fitzgerald	Dr. Sebastian V. Wendeborn
Dr. Manuel Baca	Dr. David Lee
Professor Charles G. Shevlin	Professor Stephen B. H. Kent
W. Frank Booth	Dr. Peter Kast
Christopher Claiborne	Benjamin F. Cravatt
Martine Reymond	Professor Jan Sjövall
Thomas Lebhar	Michael Remener

The support of Mass Consortium Corporation [(619)677-9432] is acknowledged.

I also appreciate the support of Professors Richard A. Lerner, William H. Beers, and K. C. Nicolaou; they represent a rare combination of leadership, management, and scientific innovation from which I continue to learn.

The most important influence has been my wife, Milena, whose intellect and enthusiasm have been inspirational in the writing of this book.

Historical Tabulation of Mass Spectrometry Developments in Biochemical Analysis

Investigator(s)	Year	Contribution(s)
J. J. Thompson	1912	First mass spectrometer
Dempster	1918	Developed electron ionization and magnetic focusing
Aston	1919	Atomic weights using MS
Stephens	1946	Time-of-flight mass analysis
Nier-Johnson	1952	Double-focusing instruments
Paul, Steinwedel, Raether, Reinhard, and von Zahn	1953–1958	Quadrupole analyzers
Wiley and McLaren	1955	Advanced the time-of-flight mass analyzer design
Gohlke and McLafferty	1956	Gas chromatography–mass spectrometry (GC–MS)
Beynon	1956	High-resolution MS
Biemann, Seibl, and Gapp	1959	Peptide sequencing
Hipple, Sommer, and Wabschall	1965	Ion cyclotron resonance
Munson and Field	1966	Chemical ionization
McLafferty and Jennings	1967	Tandem mass spectrometry
Dole, Mack, Hines, Mobley Ferguson, and Alice	1968	Electrospray ionization
Becky	1969	Field desorption MS of organic molecules
McLafferty	1973	Liquid chromatography–mass spectrometry (LC–MS)

MacFarlane, Skowronski, and Torgerson	1974	Plasma desorption MS
Comisarow and Marshall	1974	Fourier transform ion cyclotron resonance MS
Barber, Bordoli, Sedgwick, and Tyler; Surman and Vickerman	1981	Fast atom bombardment MS
Blakely and Vestal	1983	Thermospray MS
Yamashita and Fenn; Aleksandrov, Gall, Krasnov, Nikolaev, Pavlenko, Shkurov, Dokl, Baram, and Garacher	1984	Application of electrospray ionization to macromolecules
Karas, Bachmann, Bahr, and Hillenkamp; Tanaka, Waki, Ido, Akita, Yoshida, and Yoshida	1987–1988	Matrix-assisted laser desorption/ ionization (MALDI)-MS
Henry, Williams, Wang, McLafferty, Shabanowitz, and Hunt	1989	Electrospray and ion cyclotron resonance MS
Katta and Chait	1990	Observation of protein conformational changes with electrospray MS
Wilm	1991	Microelectrospray
Henion and Ganem; Chait and Katta	1991	Noncovalent complexes observed with electrospray MS
Chait, Wang, Beavis, and Kent	1993	Protein ladder sequencing with MALDI-MS
Pieles, Zurcher, Schar, and Moser; Fitzgerald, Zhu, and Smith	1993	Oligonucleotide sequencing with MALDI-MS

Introduction

What is now proved was once only imagin'd.—William Blake

Mass spectrometry is emerging as an important tool in biochemical research. The scientists who make up the preceding chronology have developed and established this technique into what it is today, a highly sensitive tool that is capable of analyzing small and large molecules. What follows is a descriptive exploration of the present state of mass spectrometry, beginning with the fundamentals of the instrumentation and its capabilities.

Advances in chemical technology have been the engine powering the biotechnology industry. Analytical chemists have added fresh impetus to bioresearch with two new mass spectrometry ionization tools, electrospray and matrix-assisted laser desorption/ionization (MALDI). Commercial availability of these instruments has made routine the analysis of compounds including proteins, peptides, carbohydrates, oligonucleotides, natural products, and drug metabolites, offering picomole to femtomole sensitivity and enabling the direct analysis of biological fluids with a minimum amount of sample preparation. Mass spectrometry allows for the analysis of these small and large biomolecules through "mild" desorption and ionization methods. Their utility now extends beyond simple molecular weight characterization. Noncovalent interactions, protein and peptide sequencing, DNA sequencing, protein folding, *in vitro* drug analysis, and drug discovery are among the areas to which mass spectrometry is being applied.

What Is a Mass Spectrometer and What Does It Do?

A mass spectrometer is an analytical device that determines the molecular weight of chemical compounds by separating molecular ions according to their mass-to-charge ratio (m/z). The ions are generated by inducing either the loss or the gain of a charge (e.g., electron ejection, protonation, or deprotonation). Once the ions are formed they can be separated accord-

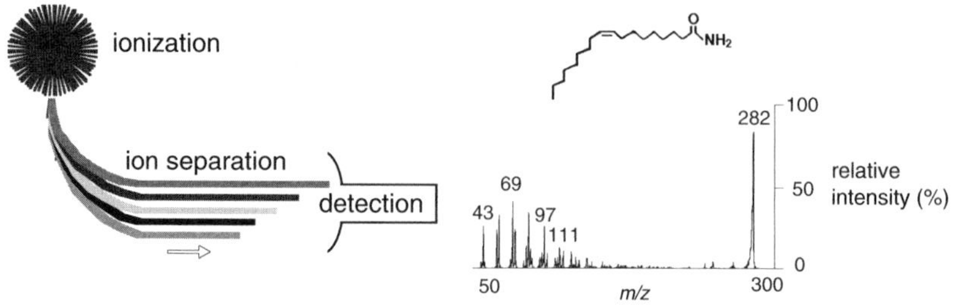

Figure I.1 Illustration of the mass analysis process and a mass spectrum of a *cis*-9,10-octadecenoamide. The original species has mass $m = 281$. Ionization adds a proton, forming a positive ion with mass-to-charge ratio $m/z = 282$, which appears as the main peak in the spectrum. The smaller ions arise from fragments of the original ion.

ing to m/z and finally detected. The result of ionization, ion separation, and detection is a mass spectrum that can provide molecular weight or even structural information (Figure I.1).

For descriptive purposes, an analogy can be drawn between a mass spectrometer and an optical spectrophotometer (Figure I.2). In the latter, light is separated into its various wavelength components by a prism and then detected with an optical receptor (such as an eye). Analogously, a

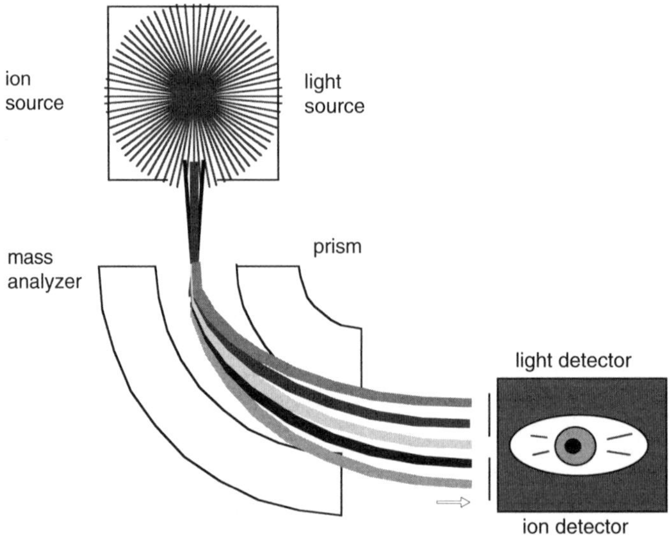

Figure I.2 Analogy between mass analysis and the dispersion of light.

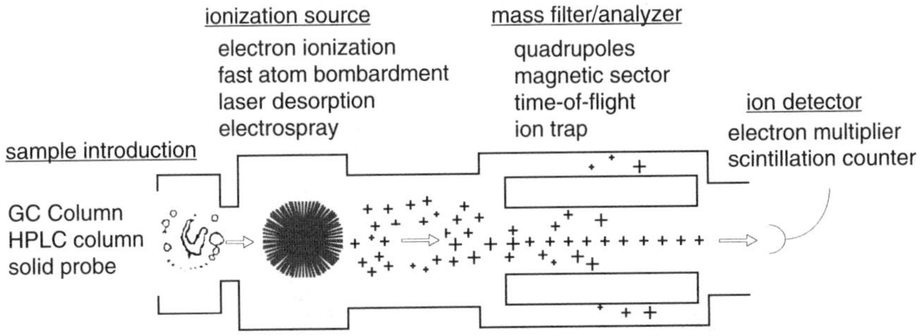

Figure I.3 Components of a mass spectrometer.

mass spectrometer contains an ion source that generates ions, a mass analyzer, which separates the ions according to their mass-to-charge ratio, and an ion detector.

Figure I.3 is an illustration of the basic components of a mass spectrometer. Once the sample is introduced into the instrument it undergoes ionization in the ionization source. The charged molecules are then electrostatically propelled into the mass analyzer/filter, which separates the ions according to their mass-to-charge ratio (*m*/*z*). The detector signal is then transferred to a computer, which stores and processes the information.

Chapter 1 focuses on how mass spectrometers transport both large and small biomolecules into the gas phase and achieve ionization. These processes are needed for the efficient and productive use of mass spectrometry in solving bioanalytical problems. Chapter 2 details essential functions of the mass analyzer and the ion detector. Chapter 3 describes how the new techniques can best be used and answers some common questions. Chapter 4 describes peptide and protein analysis and Chapter 5 describes carbohydrate, oligonucleotide, and small molecule analysis. Chapter 6 presents specific applications of how this technology is being used.

CHAPTER 1

Ion Sources and Sample Introduction

This chapter describes sample introduction and ionization. It includes common ionization techniques, such as electrospray and matrix-assisted laser desorption/ionization, and the advantages and disadvantages of each for biomolecule analysis. It also includes a brief description of the vacuum system and calibration methods.

Sample Introduction

The sample inlet is the interface between the sample and the mass spectrometer (Figure 1.1). In order to analyze by mass spectrometry, a sample at atmospheric pressure must be introduced into the instrument such that the vacuum within remains relatively unchanged. A sample can be introduced several ways, the most common being with a direct insertion probe, or by infusion through a capillary column.

The use of an insertion probe (Figure 1.2) is straightforward and involves a very simple sample introduction procedure. The sample is placed on a probe which is then inserted, usually through a vacuum lock, into the ionization region of the mass spectrometer. The sample can then be heated to facilitate thermal desorption or undergo any number of high-energy desorption processes used to achieve vaporation and ionization.

Capillary infusion is often used because it can efficiently introduce small quantities of a sample into a mass spectrometer without destroying the vacuum. Capillary columns are routinely used to interface the ionization source of a mass spectrometer with other separation techniques including gas chromatography (GC) and liquid chromatography (LC). Gas chromatography and liquid chromatography can serve to separate a solution into its different components prior to mass analysis. In gas chromatography

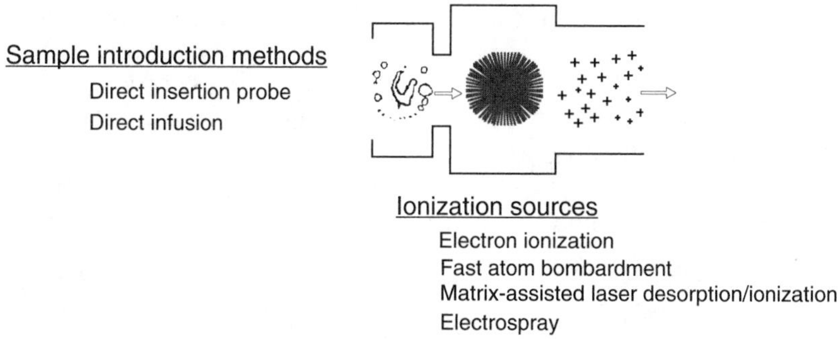

Figure 1.1 Introduction and ionization components of a mass spectrometer.

separation of sample components occurs within a capillary column. As the vaporized sample exits the gas chromatograph, it is directly introduced into the mass spectrometer. The extra information that could be obtained by adding mass spectrometry to gas chromatography was also a strong motivating factor in interfacing liquid chromatography. Prior to the 1980s, interfac-

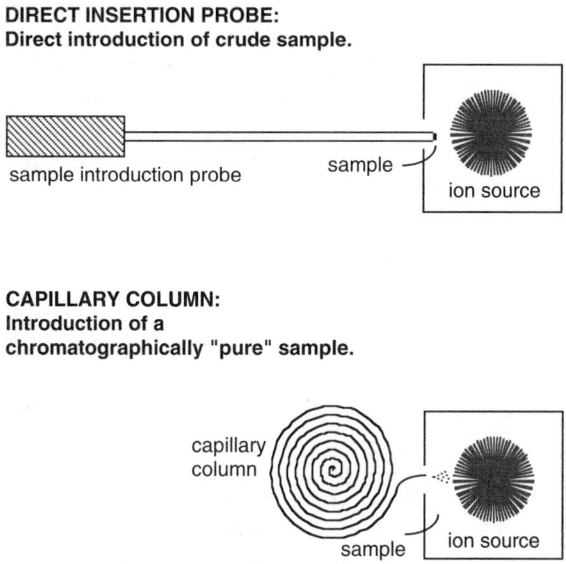

Figure 1.2 Samples are often introduced using a direct insertion probe or a capillary column. The probe and capillary carry the sample into the vacuum of the mass spectrometer. Once inside the mass spectrometer, the sample is exposed to the ionization source.

ing liquid chromatography with the available ionization techniques was difficult because of the low sample concentrations and relatively high flow rates of liquid chromatography. However, new ionization techniques were developed that now allow liquid chromatography–mass spectrometry (LC–MS) to be routinely performed (an example of LC–MS data is shown in Figure 1.3). When mass spectrometry is used as a detector for these chromatography techniques the extra information that mass analysis provides can be invaluable for sample identification.

Ionization Techniques

The ion source has undergone dramatic changes in the recent past, allowing for quick and easy analyses that previously required laborious sample preparation or were simply not possible. The sensitivity and mass range offered with both matrix-assisted laser desorption/ionization (MALDI) and electrospray ionization (ESI) (described in this section) are making these ionization techniques the methods of choice. The most common ionization sources used in biomass analysis today are listed in Table 1.1.

Electron ionization (EI) was the primary ionization source for mass analysis until the 1980s, limiting the chemist to small molecules well below the mass range of common bioorganic compounds. This limitation motivated scientists such as John B. Fenn, Franz Hillenkamp, Michael Karas,

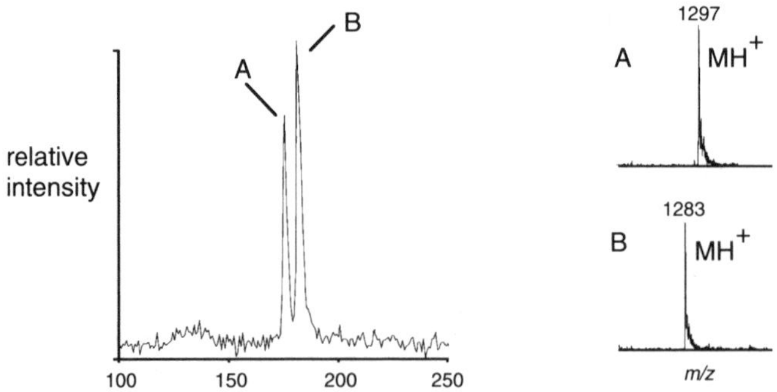

Figure 1.3 Liquid chromatography–mass spectrometry (LC–MS) ion chromatogram and the corresponding electrospray mass spectra of two peptides. Gas chromatography–mass spectrometry (GC–MS) produces results in much the same way as LC–MS; however, with GC–MS an electron ionization source is used. These ionization sources will be described in the following section on ionization techniques.

TABLE 1.1
Methods for Ionization of Bioorganic Compounds

Ionization technique	Acronym	Means of ionization
Electron ionization	EI	Electron beam/electron transfer
Fast atom/ion bombardment	FAB	Ion desorption/proton transfer
Matrix-assisted laser desorption/ ionization	MALDI	Photon absorption/proton transfer
Electrospray ionization	ESI	Evaporation of charged droplets

and Michael Barber to develop the techniques for biomolecule analysis that are now commonly known as fast atom/ion bombardment (FAB), MALDI, and ESI.

Electron Ionization

Electron ionization plays an important role in the routine analysis of small molecules. In fact, databases containing the electron ionization mass spectra of over 100,000 compounds currently exist and are used daily by thousands of chemists. These databases, combined with current computer storage capacity and searching algorithms, allow for rapid comparison with known mass spectra, thus facilitating the structural determination of small molecules.

The electron ionization technique is straightforward. The sample must be delivered as a gas—a process usually accomplished by thermal desorption or "boiling off" the sample from the probe. The high vacuum of the mass spectrometer, together with heating, can facilitate this thermal desorption process. Once in the gas phase, the compound passes into an electron ionization region (Figure 1.4) where it interacts with an electron beam, resulting in electron ejection and some degree of fragmentation.

The usefulness of electron ionization decreases significantly for compounds above a molecular weight of 400 Da. The requirement that the sample be thermally desorbed into the ionization source often leads to decomposition prior to vaporization. The principal problems associated with thermal desorption in electron ionization are the involatility of large molecules, thermal decomposition, and in many cases, excessive fragmentation.

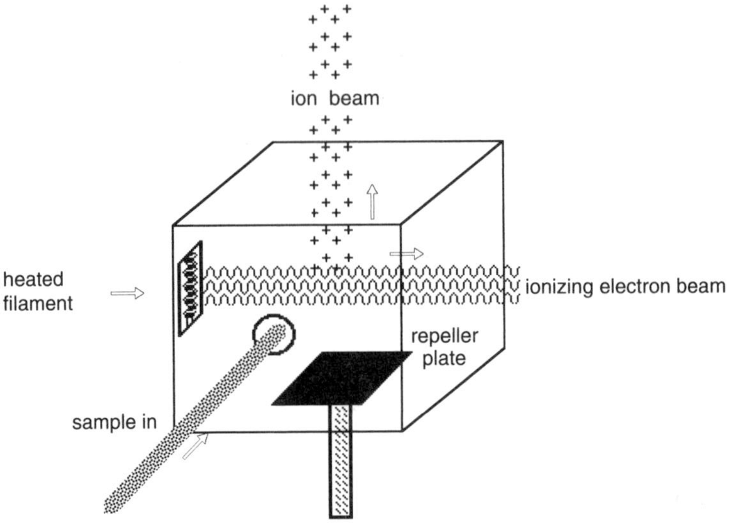

Figure 1.4 Electron ionization (EI) source.

The positive ion formation/electron ejection process in EI involves the following steps:

1. Electrons ejected from a heated filament are accelerated through an electric field at 70 V to form an electron beam.
2. The sample is thermally desorbed into this electron beam.
3. The electrons (having kinetic energy of 70 electron volts or 70 eV) transfer some of their kinetic energy to the molecule M. This transfer results in ionization (electron ejection) with usually no more than a 6 eV excess energy.

$$M + e^{-}\ (70\ \text{eV}) \rightarrow M^{+\cdot}\ (\sim 5\ \text{eV}) + 2e^{-}\ (\sim 65\ \text{eV})$$

4. Excess energy (6 eV) in the molecule leads to some degree of fragmentation.

$$M^{+\cdot} \rightarrow \text{molecular ions} + \text{fragment ions} + \text{neutral fragments}$$

In addition to electron ejection, electron ionization may also induce the negative ionization process of electron capture. Electron capture, usually 100 times less efficient than electron ejection, is most effective with compounds having a high electron affinity:

$$M + e^{-} \rightarrow M^{-}$$

Table 1.2 characterizes some of the advantages and disadvantages associated with electron ionization.

TABLE 1.2
Advantages and Disadvantages of Electron Ionization

Advantages	Disadvantages
Subpicomole to picomole sensitivity.	Limited mass range due to thermal desorption (volatility) requirement.
Availability of vast computer databases, containing over 100,000 compounds.	Possible decomposition by thermal desorption prior to vaporization.
Use of fragmentation pattern as a fingerprint with databases to identify unknowns.	Too much fragmentation, often resulting in no observable molecular ion.
Structural information obtained from fragmentation pattern.	

Fast Atom/Ion Bombardment

The FAB ionization technique (Figure 1.5) typically requires the use of a direct insertion probe for sample introduction and uses a high-energy beam of Xe atoms, Cs^+ ions, or massive glycerol-NH_4^+ clusters to sputter the sample and matrix from the probe surface. The matrix is essentially a nonvolatile solvent in which a sample is dissolved.

> **FAB matrix**–A nonvolatile liquid material that serves to constantly replenish the surface with new sample as the incident ion beam bombards this surface. The matrix also serves to minimize sample damage from the high-energy particle beam by absorbing most of the incident energy and the matrix is believed to facilitate the ionization process.

Two common matrices used with FAB are *m*-nitrobenzyl alcohol and glycerol (discussed further in Chapter 3).

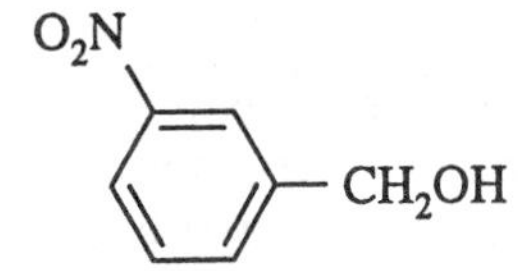

m-nitrobenzyl alcohol (NBA)

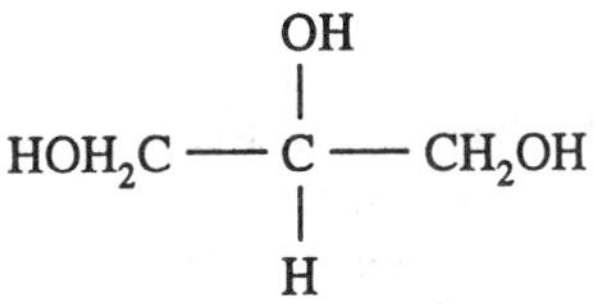

glycerol

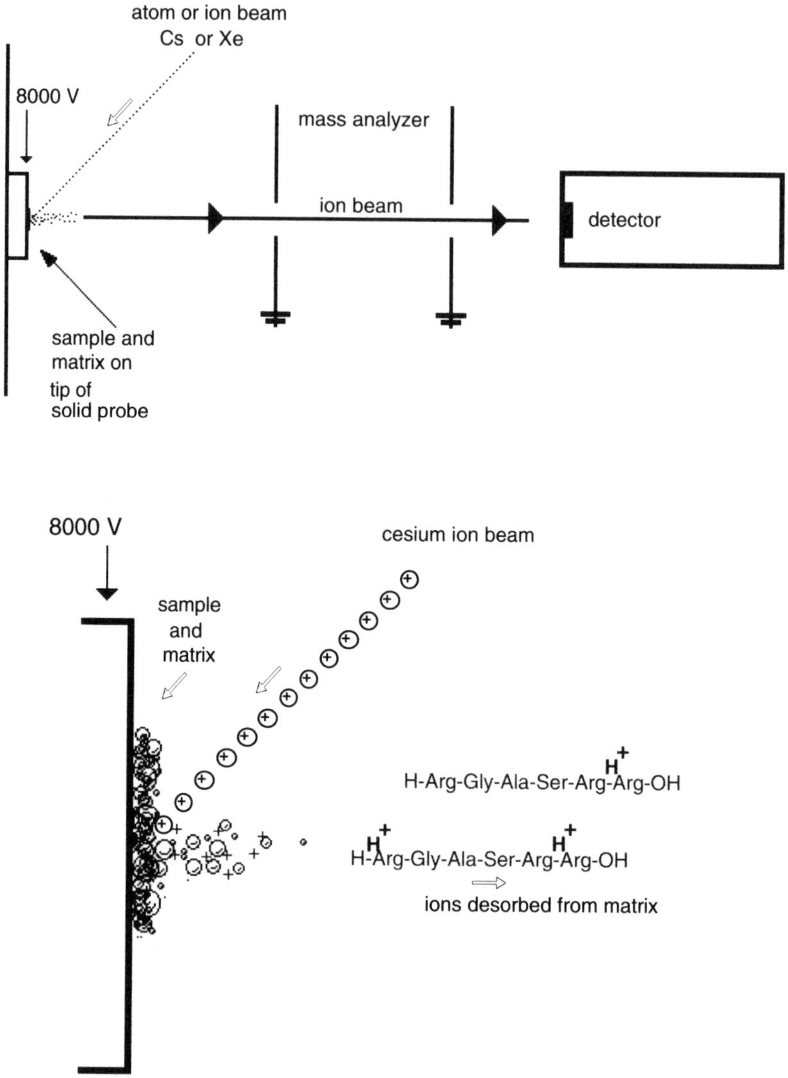

Figure 1.5 Fast atom/ion bombardment (FAB) source.

The fast atoms, ions, or clusters "splash" into the matrix and enable the sample of interest to be desorbed from the matrix solution into the gas phase. The sample may already be charged or it may become charged through reactions with surrounding molecules or ions. Charged molecules

TABLE 1.3
Advantages and Disadvantages of Fast Atom/Ion Bombardment (FAB)

Advantages	Disadvantages
Practical mass range of up to 7000 Da.	Severe drop in sensitivity at high mass.
Analyses can be performed quickly.	Relatively low sensitivity when compared to MALDI and electrospray. FAB requires high picomole to low nanomole quantities of material.
A "soft ionization" technique. The molecular ion is readily obtainable with little fragmentation.	Little fragmentation, so only a limited amount of structural information can be obtained.
Ease of adding cations to the matrix to promote molecular ion formation via cationization (e.g., $M + Na^+$, $M + K^+$...).	High background matrix peaks, confusing analysis. A significant amount of experience with the technique is necessary to be familiar with these background ions.
Ease of adaptation to high resolution instrumentation, allowing for accurate mass measurement. Matrix can be useful as a reference ion for exact mass analysis.	Requirement of solubility of sample in matrix.
Massive cluster impact (MCI) with glycerol clusters, producing multiply charged ions, making it more suitable for high-molecular-weight biopolymers. MCI is also more sensitive than conventional FAB.	Low utility for very nonpolar species that are not chargeable.

can then be propelled electrostatically to the the mass analyzer. Table 1.3 characterizes some of the advantages and disadvantages associated with FAB.

Matrix-Assisted Laser Desorption/Ionization

Matrix-assisted laser desorption/ionization–mass spectrometry (MALDI–MS) permits the analysis of high-molecular-weight compounds with high sensitivity. MALDI (Figure 1.6) is a method that allows for the ionization and transfer of a sample from a condensed phase to the gas phase in a fashion similar to FAB. The primary difference between MALDI and FAB is that while FAB uses an atom or ion beam and a liquid matrix, MALDI uses a solid matrix, and the ionizing beam is laser light. Ion

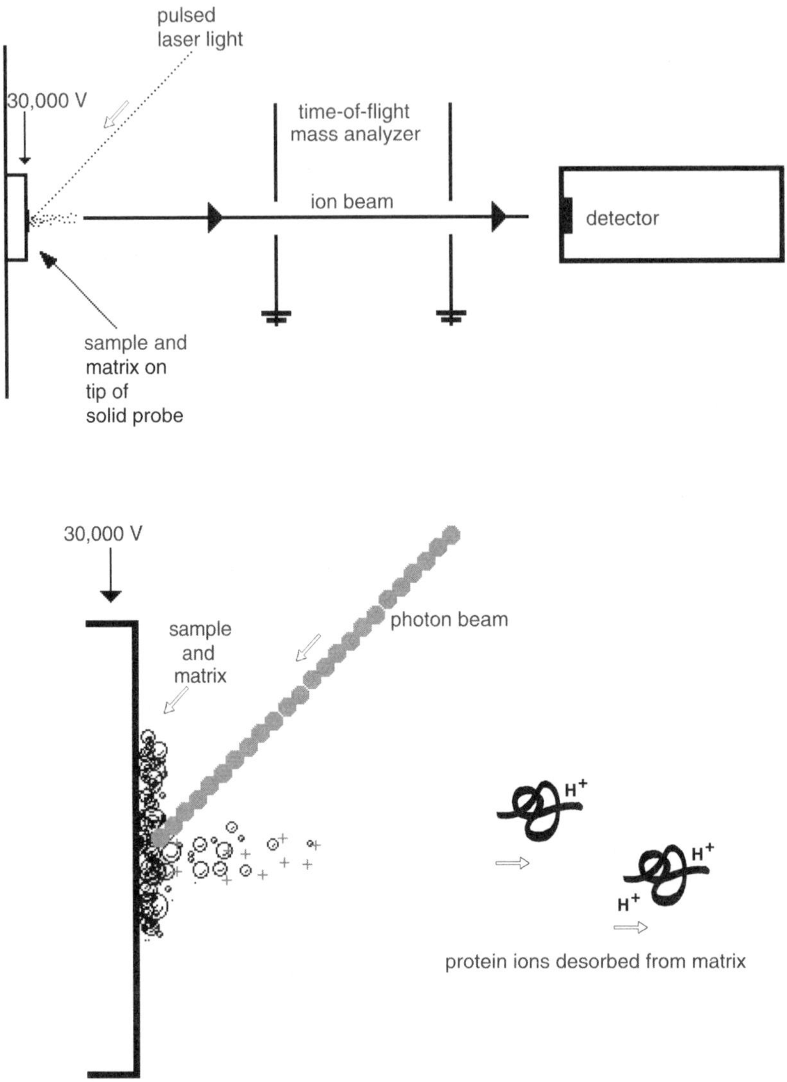

Figure 1.6 Matrix-assisted laser desorption/ionization (MALDI) source.

formation in MALDI is accomplished by directing a pulsed laser beam onto a sample suspended or dissolved in a matrix. The matrix plays a key role in this technique by absorbing the laser light energy and causing the matrix material to vaporize. (The vaporized matrix will carry some of the sample with it.) Once in the gas phase, the matrix may play a role in the ionization of the analyte molecules. The charged molecules will then be directed by electrostatic lenses from the ionization source into the mass analyzer. Uncharged molecules will often react with the matrix or other molecules to produce charged species, transferred electrostatically into the mass analyzer. Once the molecules in the sample are vaporized, time-of-flight mass analysis (Chapter 2) is often used to separate the ions according to their mass-to-charge ratio (m/z).

MALDI matrix–A nonvolatile solid material that absorbs the laser radiation resulting in the vaporization of the matrix and sample embedded in the matrix. The matrix also serves to minimize sample damage from the laser radiation by absorbing most of the incident energy and the matrix is believed to facilitate the ionization process.

The efficient and directed energy transfer during a matrix-assisted laser-induced desorption event allows for relatively small quantities of sample to be analyzed. In addition, the utility of MALDI for the analysis of heterogeneous samples makes it very attractive for the mass analysis of biological samples. Table 1.4 characterizes the advantages and disadvantages of the MALDI technique.

TABLE 1.4
Advantages and Disadvantages of Matrix-Assisted Laser Desorption/Ionization (MALDI)

Advantages	Disadvantages
Practical mass range of up to 300,000 Da. Species of much greater mass have been reported.	Low resolution (see Chapter 2). Some MALDI instruments are capable of higher resolution; however, this is only in a relatively low mass range and is accomplished at the expense of sensitivity.
Typical sensitivity on the order of low femtomole to low picomole. Reports have indicated that attomole sensitivity is possible.	Matrix background, which can be problem for compounds below a mass of 1000 Da. This background interference is highly dependent on the matrix material.
Soft ionization with little to no fragmentation observed.	Possibility of photodegradation by laser desorption/ionization.
Tolerance of salts in millimolar concentrations.	
Suitable for the analysis of complex mixtures.	

Electrospray Ionization

ESI is a method used to produce gaseous ionized molecules from a liquid solution. This is done by creating a fine spray of highly charged droplets in the presence of a strong electric field. (An illustration of the electrospray ionization process is shown in Figure 1.7.) The sample solution is sprayed from a region of a strong electric field at the tip of a metal nozzle maintained at approximately 4000 V, and the highly charged droplets are then electrostatically attracted to the mass spectrometer inlet. Either dry gas, heat, or both are applied to the droplets before they enter the vacuum of the mass spectrometer, thus causing the solvent to evaporate from the surface. As the droplet decreases in size, the electric field density on its surface increases. The mutual repulsion between like charges on this surface becomes so great that it exceeds the forces of surface tension, and ions begin to leave the droplet through what is known as a "Taylor cone" (Figure 1.8). The ions are directed into an orifice through electrostatic lenses leading to the mass analyzer.

Electrospray ionization is conducive to the formation of multiply charged molecules. This is an important feature since the mass spectrometer measures the m/z, making it possible to observe very large molecules with an instrument having a relatively small mass range. Figure 1.9 illustrates how the ions from a protein of mass 10,000 Da can be observed. Myoglobin provides another example, as seen in Figure 1.10, where each of the peaks

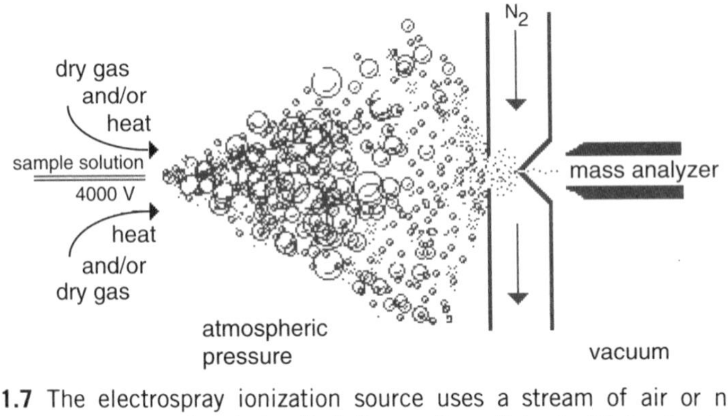

Figure 1.7 The electrospray ionization source uses a stream of air or nitrogen, heat, a vacuum, or a solvent sheath (often methanol) to facilitate desolvation of the droplets.

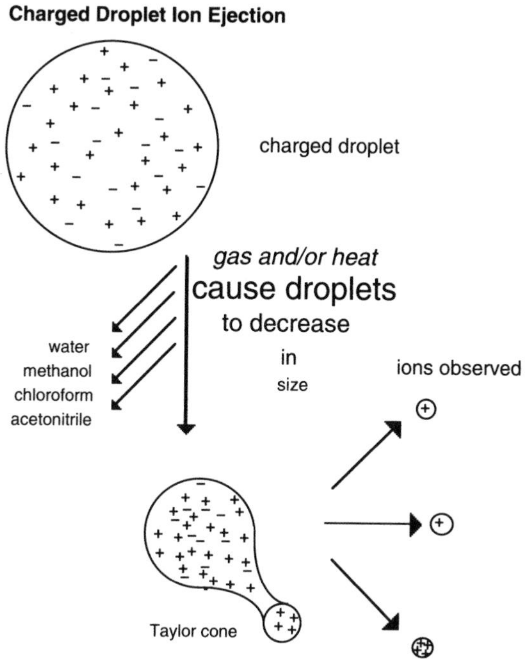

Figure 1.8 Ion formation from electrospray ionization source.

can be associated with different charge states of the molecular ion (Example 1.1).

Fortunately, computer programs available with electrospray mass spectrometers facilitate these molecular weight calculations. However, it is useful to know (at least on a rudimentary level) what the computer is calculating. A simple form of this calculation is shown here assuming p_1 and p_2 are adjacent peaks and differ by the addition of a single proton.

$$p = m/z \tag{1.1}$$

$$p_1 = (M_r + z_1)/z_1 \tag{1.2}$$

$$p_2 = \{M_r + (z_1 - 1)\}/(z_1 - 1) \tag{1.3}$$

p = a peak in the mass spectrum
m = total mass of an ion
z = total charge
M_r = average mass of sample

p_1 = m/z value for peak one
p_2 = m/z value for peak two
z_1 = charge on peak p_1

These equations can be solved for the two unknowns, M_r and z_1.

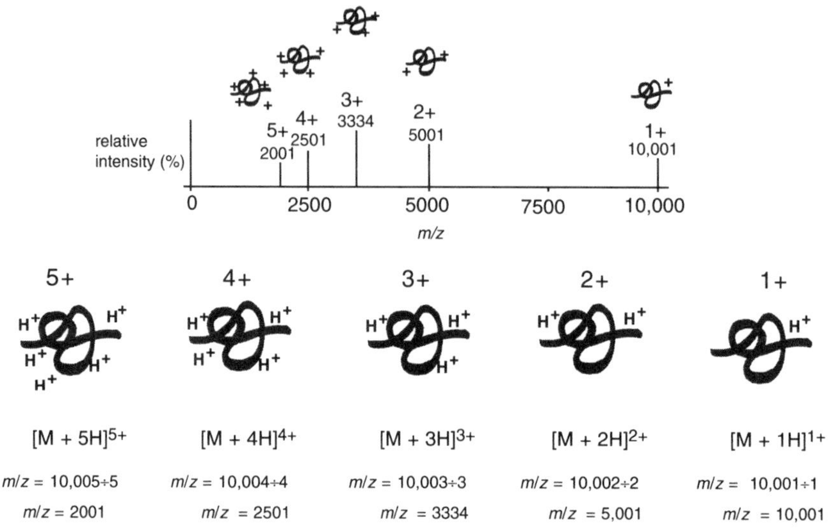

Figure 1.9 The same protein with a molecular weight of 10,000 contains 5, 4, 3, 2, and 1 charges. The mass spectrometer detects the protein ions at *m/z* = 2001, 2501, 3334, 5001, and 10,001, respectively.

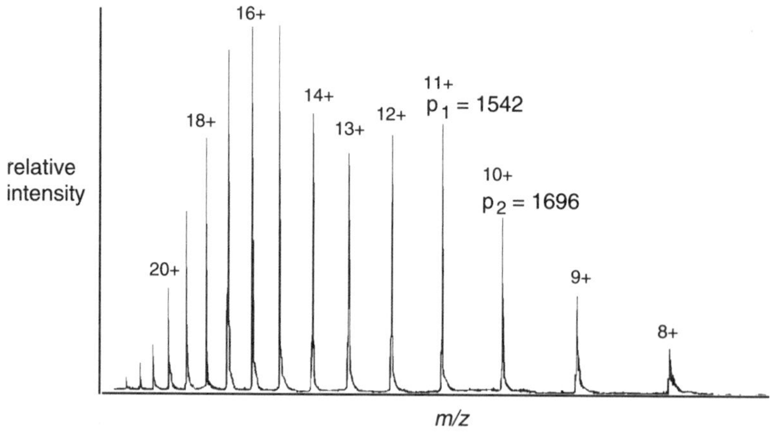

Figure 1.10 The electrospray mass spectrum of myoglobin. The different peaks represent different charge states of the *same* molecule, myoglobin. The molecular weight of the protein can be determined using Equations 1.1–1.3 and the *m/z* of p_1 and p_2.

Example 1.1: A given protein of mass 10,000 Da would generate a number of ion signals depending on the type of ion source used. However, regardless of the ion source, the mass-to-charge ratio (*m/z*) is what is being measured. A protein is shown in Figure 1.9 with one charge, two charges, ... up to five charges. The mass spectrum is also shown where the mass of the protein remains the same, yet the *m/z* ratio varies depending on the number of charges on the protein. Protein ionization is usually the result of protonation (as in Figure 1.9), which increases the mass by the number of protons added. This measurement of *m/z* applies equally for any mechanism of molecular ionization, including the addition or ejection of charge-carrying species other than protons (e.g., Na^+, Cs^+, electrons), resulting in a positively *or* negatively charged molecular ion. Multiple negative charging is typical for oligonucleotides.

For the peaks in the electrospray mass spectrum of myoglobin shown in Figure 1.10, p_1=1542 and p_2=1696.

$$1542\ z_1 = M_r + z_1 \tag{1.4}$$

$$1696\ (z_1 - 1) = M_r + (z_1 - 1) \tag{1.5}$$

Solving the two equations:

$$M_r = 16{,}951 \text{ Da and } z_1 = 11$$

Some of the advantages and disadvantages associated with electrospray are listed in Table 1.5. A general comparison for all of the ionization techniques mentioned here is presented in the Appendix.

Mechanisms of Ionization

The ion sources just described produce ions either by ionizing a neutral molecule through electron ejection, electron capture, protonation, cationization (e.g., M + Na^+), or deprotonation; or by the transfer of a charged molecule from a condensed phase to the gas phase. The present section briefly summarizes each of these ionization mechanisms.

Protonation

Protonation involves the addition of a proton to a molecule to produce a net positive charge of 1+ for every proton added. Protonation of a peptide

TABLE 1.5
Advantages and Disadvantages of Electrospray Ionization (ESI)

Advantages	Disadvantages
Practical mass range of up to 70,000 Da.	Low salt tolerance.
Good sensitivity. Femtomole to low picomole sensitivity is typical.	Difficulty in cleaning overly contaminated instrument due to high sensitivity for certain compounds.
Softest ionization. Capability of observing biologically native noncovalent interactions.	Low tolerance for mixtures. Simultaneous mixture analysis can be poor. The purity of the sample is important.
Easily adaptable to microbore liquid chromatography. Capable of directly analyzing LC effluent at a flow rate of one milliliter per minute.	Multiple charging, which can be confusing, especially with mixture analysis.
No matrix interference.	
Easy adaptability to triple quadrupole analysis, conducive to structural analysis.	
Multiple charging, allowing for the analysis of high-mass ions with a relatively low *m/z* range instrument.	
Multiple charging, giving better mass accuracy through averaging.	

is shown in Scheme 1.1, with the corresponding mass spectrum obtained in FAB. Charges tend to reside on the more basic residues of the molecule.

Cationization

Cationization involves the noncovalent addition of a positively charged ion to a neutral molecule, resulting in a charged complex. While protonation can be thought of as cationization, the term cationization is more commonly used for the addition of a cation adduct other than a proton. Also, this ionization mechanism is especially useful with molecules that are not stable to protonation. Protonation adds a charge onto a molecule, yet because of the covalent nature of proton binding the charge can be delocalized from the proton onto the molecule. This charge delocalization can destabilize

$$M + H^+ \longrightarrow MH^+$$

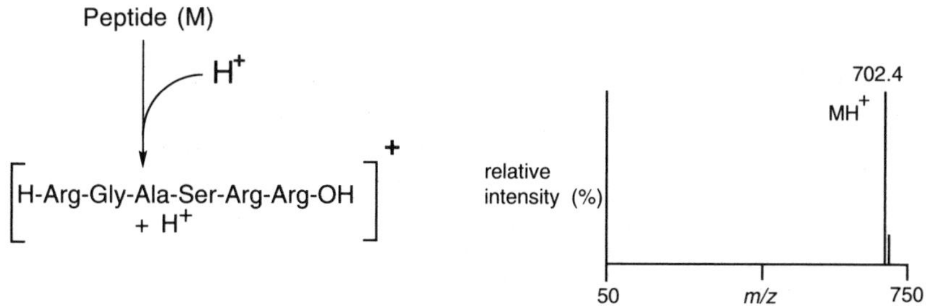

Scheme 1.1 An example of a mass spectrum obtained via protonation. Protonation is commonly achieved via fast atom/ion bombardment (FAB), matrix-assisted laser desorption/ionization (MALDI), and electrospray ionization (ESI).

the molecular ion, resulting in fragmentation. The binding of cations other than protons (e.g., alkali, ammonium) to a molecule is usually less covalent in nature, and the charge remains localized on the cation, thereby minimizing charge delocalization and fragmentation. Cationization is often utilized in the desorption and electrospray ionization techniques to produce a stable molecular cation. Carbohydrates are excellent candidates for this ionization mechanism, with Na^+ a common cation adduct (Scheme 1.2).

$$M + Cation^+ \longrightarrow MCation^+$$

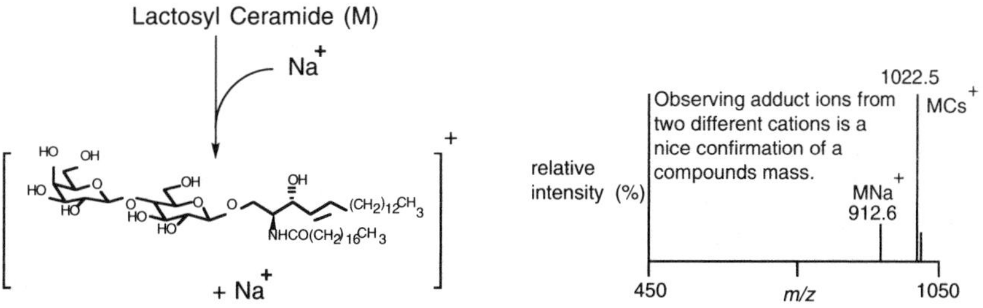

Scheme 1.2 An example of a mass spectrum obtained via cationization. Cationization is commonly achieved via fast atom/ion bombardment (FAB), matrix-assisted laser desorption/ionization (MALDI), and electrospray ionization (ESI).

Deprotonation

Deprotonation is the ejection of a proton from a molecule, resulting in a net negative charge of 1− for each proton ejected. This mechanism of ionization is very useful for acidic species, including phenols, carboxylic acids, and sulfonic acids. With deprotonation, the net positive charge of 1− is achieved through the removal of a proton or even multiple negatively charged species, as observed in the electrospray of oligonucleotides. The mass spectrum of the carbohydrate, sialic acid, is shown in Scheme 1.3.

Transfer of a Charged Molecule into the Gas Phase

Many compounds are already charged in solution, and it is only necessary to transfer them into the gas phase. This usually occurs by desorbing or ejecting the charged species from the condensed phase into the gas phase (Scheme 1.4). Once in the gas phase the molecule can then be electrostatically deflected into the mass analyzer.

Electron Ejection

Electron ejection is observed most commonly with EI sources, and usually performed on relatively nonpolar low-molecular-weight compounds. As its name implies, electron ejection involves the ejection of an electron to produce a net positive charge of 1+. The electron ejection of anthracene is shown in Scheme 1.5, with the corresponding mass spectrum obtained with electron ionization.

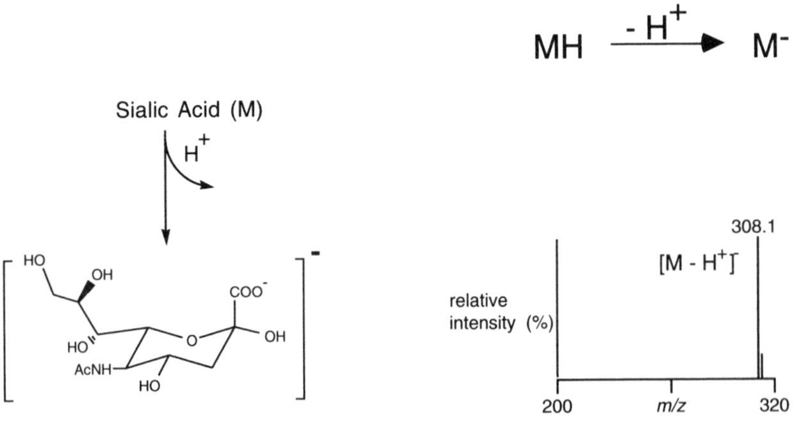

Scheme 1.3 An example of a mass spectrum obtained via deprotonation. Deprotonation is commonly achieved via fast atom/ion bombardment (FAB), matrix-assisted laser desorption/ionization (MALDI), and electrospray ionization (ESI).

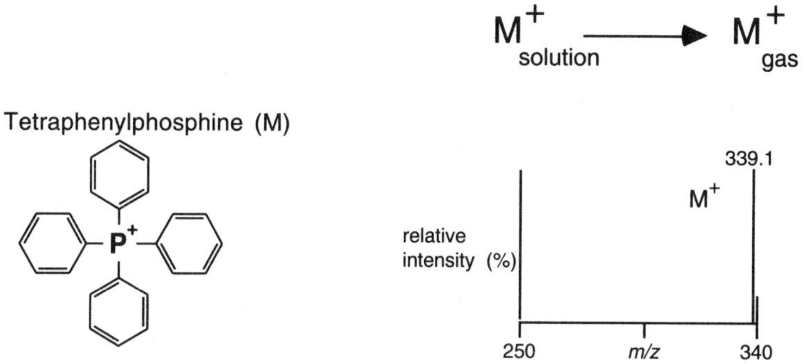

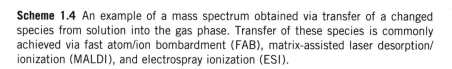

Scheme 1.4 An example of a mass spectrum obtained via transfer of a changed species from solution into the gas phase. Transfer of these species is commonly achieved via fast atom/ion bombardment (FAB), matrix-assisted laser desorption/ionization (MALDI), and electrospray ionization (ESI).

Electron Capture

Electron capture involves the absorption or capture of an electron to produce a net negative charge of 1−. It is a mechanism of ionization primarily observed for molecules with a high electron affinity (e.g., halogenated compounds such as hexachlorobenzene) and most often with electron ionization sources (Scheme 1.6). Electron capture is also observed with the FAB and MALDI particle desorption ionization techniques.

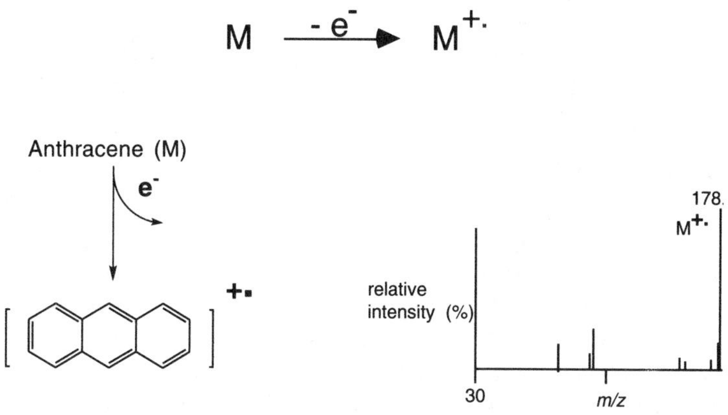

Scheme 1.5 An example of a mass spectrum obtained via electron ejection. Electron ejection is commonly achieved via electron ionization (EI). FAB and MALDI can also induce electron ejection.

$$M \xrightarrow{+e^-} M^{+\cdot}$$

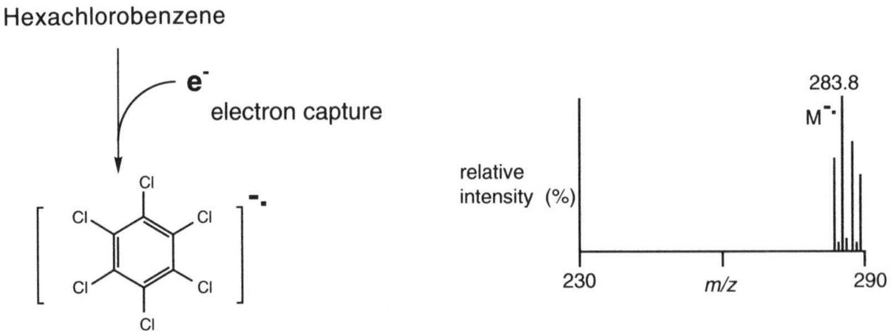

Scheme 1.6 An example of a mass spectrum obtained via electron capture. Electron capture is commonly achieved via electron ionization (EI).

Calibration

Mass accuracy is one of the most important aspects of the data obtained from a mass spectrometer. In order to maintain high accuracy the mass spectrometer must be calibrated on a regular basis or at least some reference compound must be checked to determine if the instrument has not lost accuracy. The process of correcting a mass spectrometer for better accuracy is known as calibrating. Calibrating mass spectrometers requires compounds that produce a reliable source of ions with a known mass. When this known compound is analyzed, one can readily determine if the instrument is running within acceptable error limits. If not, a simple adjustment will allow the instrument to run accurately again. This section briefly discusses calibration for FAB, MALDI, and electrospray ionization and the compounds used to perform calibrations.

Calibration for FAB

The salt of cesium iodide (CsI) is probably the most common material used for FAB calibration. CsI readily forms cluster ions 260 mass units apart, up to and greater than 10,000 *m/z*. Thus, a reliable signal may be generated over a broad range. Also, both cesium and iodide have only one isotope, so there are no additional isotope peaks to complicate the spectra. In addition, CsI holds a consistent signal strength, and when stored in the dark, will last months to years. Other calibration compounds (e.g., polyethylene glycols and other salt clusters, NaI, CsF) are also used. Figure 1.11 shows the CsI cluster ions used as reference ions in both FAB and electrospray.

Figure 1.12 illustrates a typical spectrum obtained with *m*-nitrobenzyl alcohol (NBA). An added benefit of using these cation adducts is that they

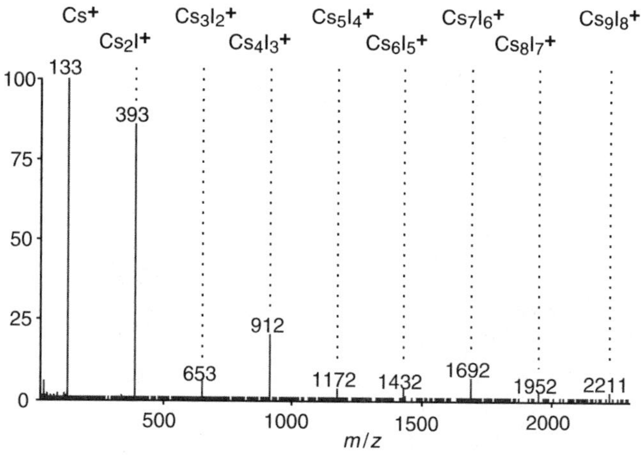

Figure 1.11 Electrospray mass analysis of CsI clusters. A similar spectrum is observed with FAB analysis.

can provide reference ions for exact mass analysis. The NBA/CsI mixture can produce ions well above 1000 *m/z* (see Appendix).

One trick that can be helpful both for obtaining reference peaks and for producing cationization is adding a small amount of CsI or NaI dissolved in an aqueous solution to the NBA matrix. Compared to CsI, a much smaller amount of NaI is needed, but NaI does not generate reference ion clusters with as high a molecular weight as CsI.

Calibration for MALDI

The calibration of a MALDI mass spectrometer (MALDI–time-of-flight) only requires a couple of points. For instance, Na^+ (*m/z* 23) and a small peptide or even matrix ions can be used in the low–mass region, while the singly and doubly charged ions of the proteins bovine insulin (MW = 5734 Da), horse myoglobin (MW = 16,951 Da), and cytochrome

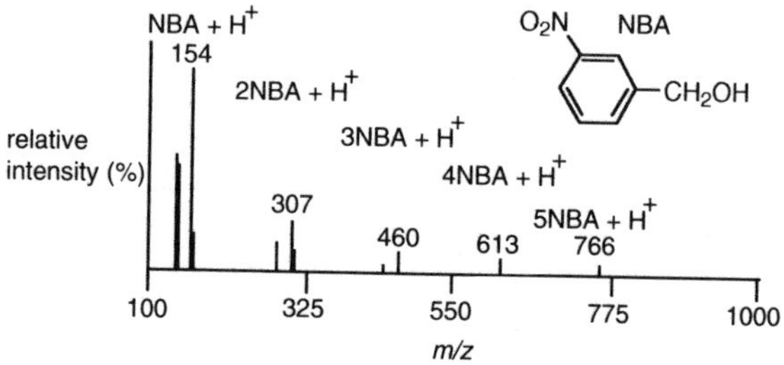

Figure 1.12 FAB mass spectrum of *m*-nitrobenzyl alcohol (NBA).

c (MW = 12,361 Da) are useful in the high–mass region. To ensure accurate data acquisition, however, it is prudent to have an internal standard mixed in with the compound of interest; often one of the proteins shown in Table 1.6 provides an adequate internal reference.

Calibration for Electrospray

Polypropylene glycols (PPGs, Figure 1.13) are probably the most common compounds used to calibrate electrospray instruments. CsI dissolved in methanol is also very useful (Figure 1.12) because it has the advantage of being monoisotopic and it has only one reference ion every 260 mass units. The PPGs generate many ions, so care must be taken not to calibrate using the wrong peak.

Vacuum in the Mass Spectrometer

A common requirement of all mass spectrometers is a vacuum. A vacuum is necessary to permit ions to reach the detector without colliding with other gaseous molecules (Figure 1.14). Such collisions would reduce the resolution and sensitivity of the instrument by increasing the kinetic energy distribution of the ion, thus inducing fragmentation, or preventing the ions from reaching the detector.

Coupling any sample source to a mass spectrometer requires that the sample (at atmospheric pressure, 760 Torr) be transferred into a region of high vacuum ($\sim 10^{-6}$ Torr) without compromising the latter. The billion-fold difference in pressure between the atmosphere and the high vacuum was one of the first problems faced by the originators of mass spectrometry.

Various mechanical configurations are now used to maintain the vacuum in a mass spectrometer while introducing the sample into the vacuum chamber. One method is to introduce the sample in small quantities through a capillary column (e.g., gas chromatography) or through a small orifice directly into the instrument (Figure 1.15). A second method is to initially evacuate the sample chamber through a vacuum lock (Figure 1.16). Once a moderate vacuum is achieved (10^{-3} Torr) using the vacuum lock in a prechamber, the sample can be introduced into the main vacuum chamber on the mass spectrometer. This method is usually employed with the direct insertion probes used with MALDI and FAB.

Vacuum Systems

A mass spectrometer is shown in Figure 1.17 with three alternative pumping systems. All three systems are capable of producing a very high vacuum, and are all backed by a mechanical pump. The mechanical pump serves as a general workhorse for most mass spectrometers and allows for

TABLE 1.6
Common MALDI Calibration Compounds and Their Ionic Mass-to-Charge (*m/z*) Ratios

Calibrants	MH^+	$2MH^+$	MH_2^{2+}	MH_3^{3+}
Matrices				
Gentisic acid (DHB)	155.03 (mono)	309.06 (mono)		
Sinapic acid (SA)	225.08 (mono)	449.14 (mono)		
α-CN-4-OH-cinnamic acid	190.05 (mono)	379.09 (mono)		
Peptides and proteins				
ACTH (18–39)	2,465.20/2,466.73 (mono/avg)			
Insulin bovine	5,734.56 (avg)		2,867.78 (avg)	
Ubiquitin	8,565.84 (avg)		4,283.43 (avg)	
Cytochrome *c* equine	12,361.09 (avg)		6,181.05 (avg)	
apo-Myoglobin equine	6,952.47 (avg)		8,476.74 (avg)	
Trypsin bovine	23,312.54 (avg)		11,656.77 (avg)	7,771.52 (avg)
BSA	66,431. (avg)		33,216. (avg)	22,144. (avg)
BSA dimer	132,859. (avg)		66,430. (avg)	44,287. (avg)

Note Mono corresponds to the monoisotopic mass and is defined as the mass of an ion for a given empirical formula calculated using the exact mass of the most abundant isotope of each element, e.g., $C_{60}H_{122}N_{20}O_{16}S_2$; monoisotopic mass = 1442.8788 Da. Avg corresponds to the average mass and is defined as the mass of an ion for a given empirical formula, calculated using the average atomic weight, average of the isotopes, for each element, e.g., $C_{60}H_{122}N_{20}O_{16}S_2$; average mass = 1443.8857 Da.

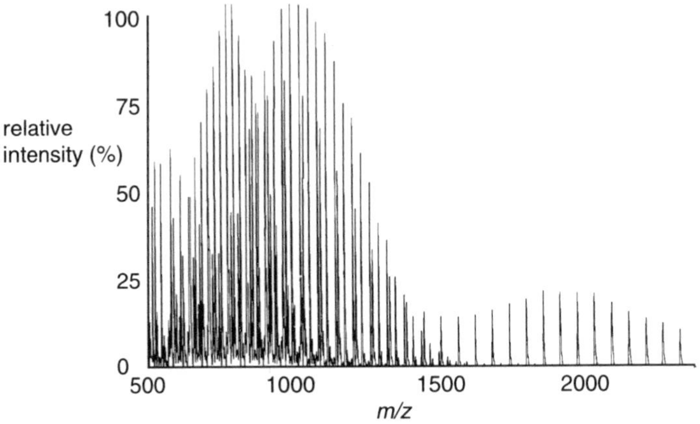

Figure 1.13 Electrospray mass spectrum of the calibration polymer polypropylene glycol.

an initial vacuum of about 10^{-3} Torr to be obtained. Once a 10^{-3} Torr vacuum is achieved, the other pumping systems can be activated to obtain pressures as low as 10^{-9} Torr. These systems are shown and described in Figures 1.18–1.21.

A well-maintained vacuum is essential to the function of a mass spectrometer. Once the vacuum is compromised, sensitivity and resolution will

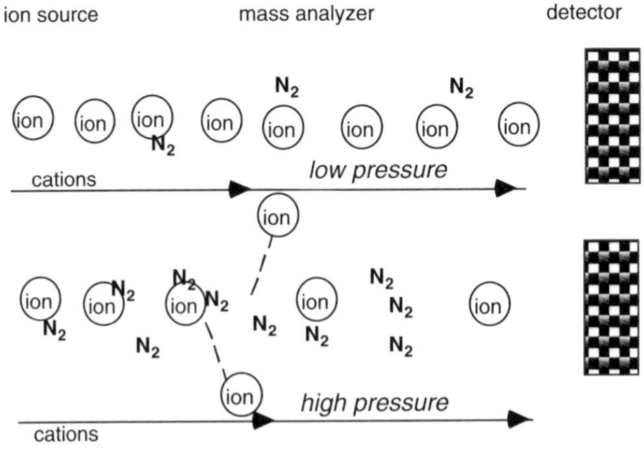

Figure 1.14 Ion–molecule interactions in a mass spectrometer will cause the sensitivity to drop and will also reduce the resolution. This effect will be increased at higher pressures because of more ion–molecule collisions.

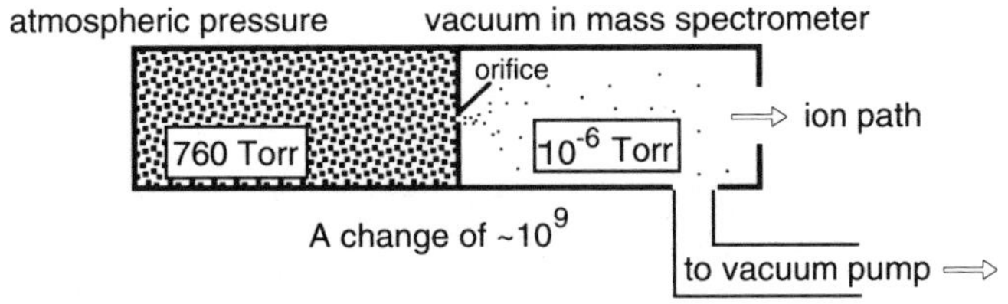

Figure 1.15 Difference in pressure between atmosphere and inside the mass spectrometer. A small orifice (or capillary column) is one means of introducing a sample from atmospheric pressure to high vacuum.

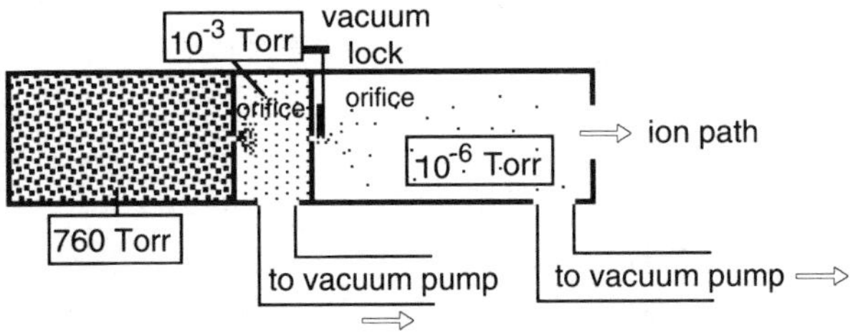

Figure 1.16 Maintaining high vacuum while introducing the sample is often accomplished using several vacuum pumps. This allows the sample to be brought to a lower pressure ($\sim10^{-3}$ Torr) before it is actually introduced into the main vacuum chamber of the mass spectrometer.

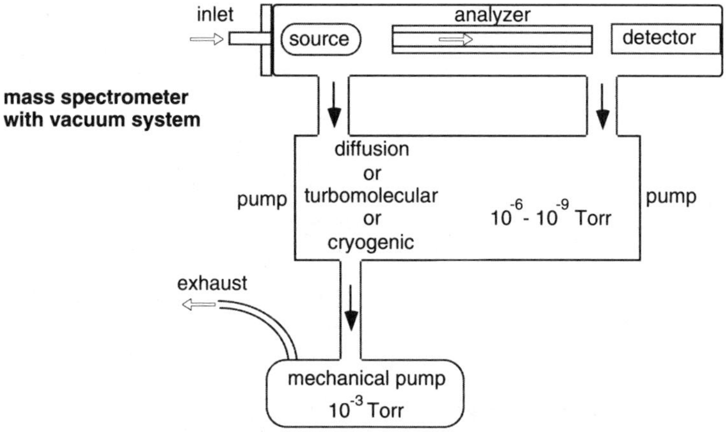

Figure 1.17 Vacuum system on a mass spectrometer.

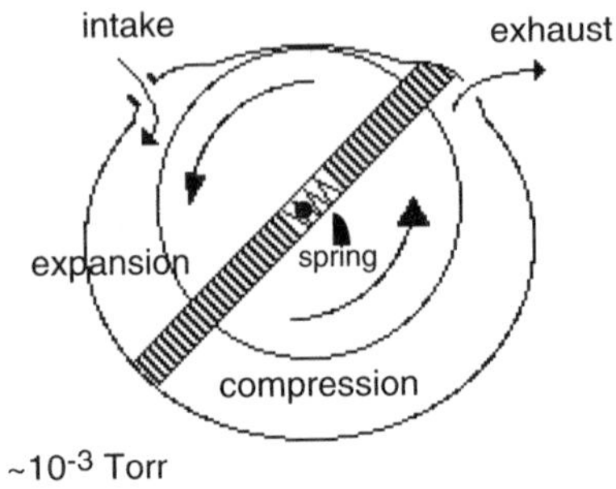
intake
exhaust
expansion
spring
compression
~10-3 Torr

Figure 1.18 Mechanical pump. High-speed vacuum pumps (diffusion and turbomolecular) are commonly backed up by mechanical pumps. Mechanical pumps produce an initial vacuum (10^{-3} Torr) for the high-vacuum pumps, whose systems usually require an initial vacuum of 10^{-3} Torr before they can operate.

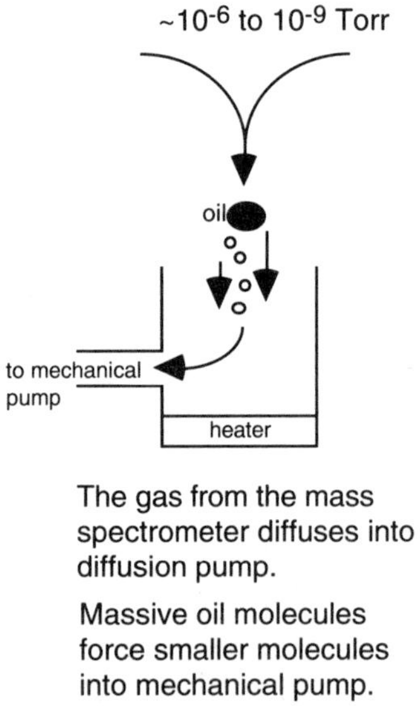
~10-6 to 10-9 Torr
oil
to mechanical pump
heater
The gas from the mass spectrometer diffuses into diffusion pump.
Massive oil molecules force smaller molecules into mechanical pump.

Figure 1.19 Diffusion pump. A diffusion pump uses an oil, often polyphenyl ether, to force molecules out of the vacuum chamber. The oil is heated at the bottom of the pump, causing it to vaporize. Water-cooled coils at the top of the pump then allow the oil to cool and condense. The big oil molecules force the smaller molecules down into the pump via momentum exchange. The smaller gaseous molecules exit through the mechanical pump (≈200 liters per second).

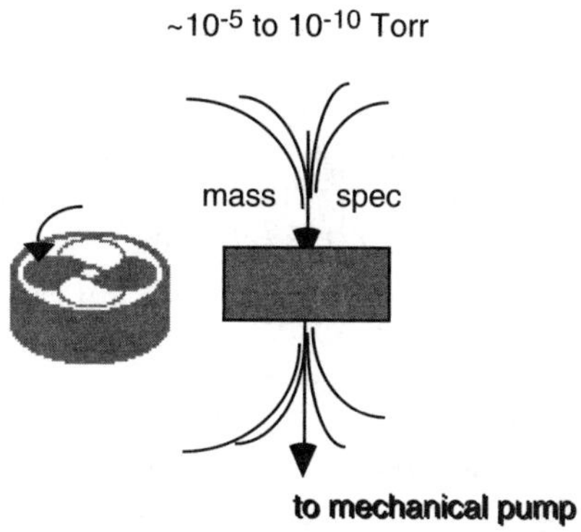

Figure 1.20 Simplified drawing of turbomolecular pumping. A turbomolecular pump is purely mechanical. Much like a jet engine or a fan, this pump also works via momentum exchange. As the molecules diffuse into the area of the pump or fan, they are forced into the mechanical pump via momentum exchange from the fins on the turbomolecular pump. The fins revolve at approximately 100,000 rpm (revolutions per minute) and produce an effective pumping capacity of about 400 liters per second.

be reduced. Also, at higher pressures the high voltages used in the instrument may discharge to ground, which can damage the instrument and/or the computer system running the instrument. Such a breakdown, which may arise from an extreme leak, is basically an implosion, and can seriously damage a mass spectrometer by destroying electrostatic lenses, coating the optics with pump oil, and damaging the detector. In general, maintaining a high vacuum is crucial to obtaining high quality spectra.

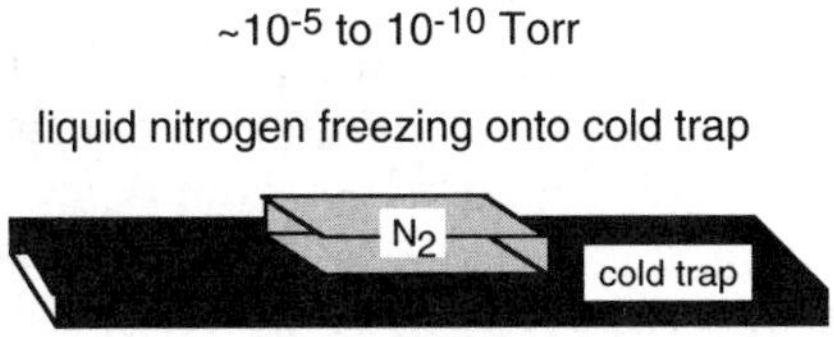

Figure 1.21 A simplified illustration of how vacuum is achieved with a cryogenic pumping system. Gases condense onto a cold surface produced by a helium compressor. A cryogenic pump operates by freezing out gases in a sealed system (the mass spectrometer). A cryogenic pumping system produces an ultra cold (~14K) surface, trapping gases on its walls or "cold finger." This type of pump is very efficient with a theoretical pumping capacity of about 100,000–200,000 liters per second!

Summary

A mass spectrometer is an analytical device which produces a signal characteristic of a species by producing, separating, and selectively detecting charged molecules. The mass spectrometer can be separated into three distinct sections: the ion source, the mass analyzer, and the detector.

A sample is introduced into the mass spectrometer by using a direct insertion probe, direct infusion, or chromatographic separation interfaced to the instrument.

The ion source produces ions by electron ejection, electron capture, cationization, deprotonation, or the transfer of a charged molecule from the condensed to the gas phase. EI was the primary ionization source for mass spectrometers until the 1980s, limiting the chemist to small molecules well below the mass range of most biological compounds. Electron ionization requires thermal desorption of the sample, which limits the utility of electron ionization to compounds below a molecular weight of 400 Da.

FAB produces a high-energy beam of Xe or Cs^+ to sputter a sample/matrix mixture into the gas phase. FAB is primarily useful for compounds in the mass range from 100 to 7000 Da and has relatively low sensitivity (high picomole to nanomole).

MALDI analysis is carried out by suspending or dissolving a sample in a solid or liquid matrix that allows for efficient and directed energy transfer during a laser–induced desorption process. MALDI–MS has proven useful for both qualitative and recently quantitative biomolecular analysis and for the analysis of heterogeneous biological samples. MALDI has a mass range routinely up to ~300,000 Da for proteins, and typical sensitivity is on the femtomole to low picomole level.

ESI is a method for ejecting ionized molecules from a solution by creating a fine spray of highly charged droplets in the presence of a strong electric field. This type of ionization is highly conducive to the formation of multiply charged molecules. Electrospray has a mass range routinely up to ~70,000 Da for proteins and typical sensitivity is on the picomole level.

An efficient pumping system is required to maintain a high vacuum while condensed phase solutions or gases at atmospheric pressure are introduced into the ionization chamber. This is accomplished with mechanical pumps used in conjunction with diffusion, turbomolecular, or cryogenic pumps.

Practice Questions

What are the similarities between the dispersion of light and mass analysis?

Why is electron ionization not useful for larger molecules?
Why isn't FAB a good technique for proteins?
Why is electrospray so useful with instruments with a small mass range?
Which ionization technique is most tolerant of salts and mixtures?
Can you give a possible explanation why electrospray is not tolerant of salts?
Which ionization technique is the most sensitive?
Which vacuum system condenses gases out of the mass spectrometer?

Other Books on Mass Spectrometry

Chapman, J. R. (1993). *Practical Organic Mass Spectrometry: A Guide for Chemical and Biochemical Analysis,* 2nd ed. Wiley, Chichester/New York.

Desiderio, D. M. (Ed.) (1991). *Mass Spectrometry of Peptides.* CRC Press, Boca Raton, FL.

Desidero, D. M. (Ed.) (1992). *Mass Spectrometry: Clinical and Biomedical Applications.* Plenum, New York.

Matsuo, T., *et al.* (Eds.) (1994). *Biological Mass Spectrometry : Present and Future* Wiley, Chichester/New York.

McCloskey, J. A. (Ed.) (1990). *Methods in Enzymology,* Vol. 193, *Mass Spectrometry.* Academic Press, San Diego.

McEwen, C. N., and Larsen, B. S. (Eds.) (1990). *Mass Spectrometry of Biological Materials.* Dekker, New York.

McLafferty, F. W., and Turecek, F. (1993). *Interpretation of Mass Spectra,* 4th ed. University Science Books, Mill Valley, CA.

Murphy, Robert C. (1993). *Mass Spectrometry of Lipids.* Plenum, New York.

NATO Advanced Study Institute on Mass Spectrometry in the Molecular Sciences (1990) *NATO ASI Series C: Mathematical and Physical Sciences,* Vol. 353, *Mass Spectrometry in the Biological Sciences: A Tutorial* (M. L. Gross, Ed.). Kluwer Academic, Dordrecht/Boston.

Russell, D. H. (Ed.) (1994). *Topics in Mass Spectrometry,* Vol. 1, *Experimental Mass Spectrometry.* Plenum, New York.

Suelter, C. H., and Watson, J. T. (Eds.) (1990). *Methods of Biochemical Analysis,* Vol. 34, *Biomedical Application of Mass Spectrometry.* Wiley, New York.

CHAPTER 2

Mass Analyzers and Ion Detectors

This chapter examines the role the mass analyzer plays in biochemical mass spectrometry. It includes common analyzer technology, such as quadrupole and time-of-flight, and the advantages and disadvantages each of these have for biomolecule analysis. Figure 2.1 illustrates a mass analyzer and its ability to separate ions. This chapter also explores how tandem mass spectrometry works and how structural information is obtained. The detection systems are also described providing a description of how ions are detected. As we begin with the mass analyzer, it is important to gain some perspective on its evolution.

The first mass analyzers, made in the early 1900s, used magnetic fields to separate ions according to their mass-to-charge ratio. Modern analyzers, whose designs include variations on the early magnetic methods, now offer high accuracy, high sensitivity, high mass range, and an ability to give structural information. Just as ionization techniques have evolved, mass analyzers have also changed to meet the demands of observing larger molecules with mass accuracy on the order of $\pm 0.01\%$ or higher. In addition, the sensitivity required (picomole to femtomole) for many biological applications has been obtained with the electrospray and MALDI ionization sources, in part due to the high transmission efficiency of the analyzers. As mentioned previously, one of the more recent developments has been the ability to obtain structural information on large biomolecules. This is achieved by inducing collisions within the mass spectrometer and then observing the fragment ions.

Mass analyzers scan or select ions over a particular *m/z* range. The key feature of all mass analyzers is their measurement of *m/z, not* the mass. This is often a point of confusion, because if an ion is multiply charged the *m/z* will be significantly less than the actual mass (e.g., an ion, $C_7H_7^{2+}$, which has a mass of 91 Da yeilds an ion at *m/z* 45.5). Multiple charging is especially common with the electrospray ionization yeilding many peaks that corre-

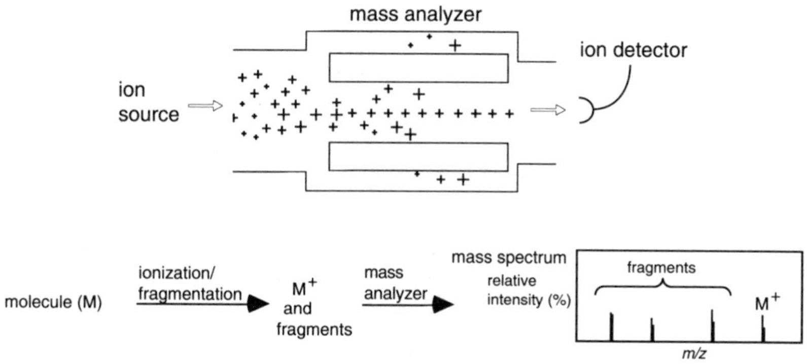

Figure 2.1 Mass analyzer and ion detection, showing measurement of the mass-to-charge ratio.

spond to the same species, which can make the spectra confusing yet at the same time has been found to be very useful (Chapter 1).

Resolution

Another common characteristic of mass analyzers is their associated resolving power, which can vary greatly from instrument to instrument. The resolving power is also identified as resolution, and is defined as follows:

> *Resolution is the ability of a mass spectrometer to distinguish between ions of different mass-to-charge ratios. Therefore, greater resolution corresponds directly to the increased ability to differentiate ions. For example, a mass spectrometer with a resolution of 500 can distinguish between ions of m/z = 500 and 501. The most common definition of resolution is given by*
>
> $$\textit{Resolution} = \mathrm{M}/\Delta\mathrm{M} = \mathrm{M}_1/(\mathrm{M}_1 - \mathrm{M}_2)$$
>
> *where the ions* M_1 *and* M_2 *are separated by a valley having a specified height above the baseline (Figure 2.2).*

Alternatively, one peak may be used to determine resolution. In that case, ΔM in the equation resolution = $M/\Delta M$ is defined as the full width at half maximum (FWHM). For example, the peak in Figure 2.2 has a mass of 500 and a FWHM of 1, therefore the resolution is $M/\Delta M = 500/1.0 = 500$.

Resolution is important because, for biopolymers, higher resolution results in better accuracy, as shown in Figure 2.3. The average mass of a

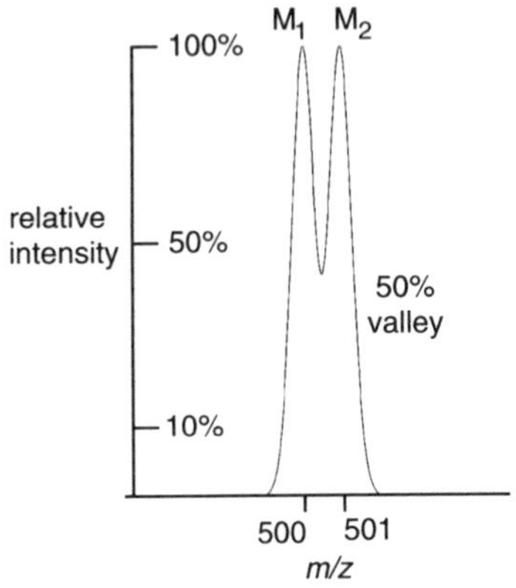

Figure 2.2 Resolution of 500 for a mass analyzer using the 50% valley definitions. To use the formula resolution = $M/\Delta M$ more precisely, it is necessary to define at what stage the two peaks, representing the two masses M_1 and M_2, are actually separated. The depth of the valley between the two peaks is used for this purpose, with 5, 10, and 50% valley definitions referring to 5% above baseline, 10% above baseline, and 50% above baseline, respectively. A 5% valley definition is a much more strict criterion of separation (resolution) than is the 50% valley definition.

molecule is calculated using the weighted average mass of all isotopes of each constituent element of the molecule, while the monoisotopic mass is calculated by using the mass of the elemental isotope having the greatest abundance for each constituent element. Many instruments cannot resolve between the isotopes and will give only the average mass, yet sometimes

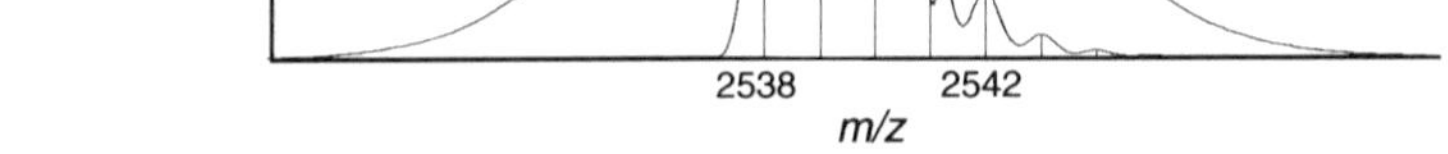

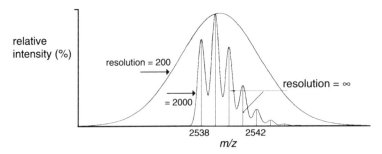

Figure 2.3 The effect of resolution upon mass accuracy. The above spectra were calculated for the same molecular formula at resolutions of 200, 2000, and infinity (∞) using the 10% valley definition. Notice that the accuracy is reduced by the uncertainty associated with identifying the center of the peak with the lower resolution.

the resolution is so low that it may be difficult to determine even the average mass with reasonable accuracy. Higher resolution allows separation of ions according to individual isotopes or simply produces a very narrow peak allowing for greater accuracy. Figure 2.3 is an example of how resolution affects the observed mass for a compound having a molecular formula (MH^+) $C_{101}H_{145}N_{34}O_{44}$.

Mass Analyzers

Instruments have variations in their capabilities that depend on their design and intended purpose. This is also true for mass spectrometers, the mass analyzer contributes to the accuracy, range, and sensitivity of an instrument. Six common types of mass analyzers are quadrupole, magnetic sector, time-of-flight, time-of-flight reflectron, quadrupole ion traps, and Fourier transform–ion cyclotron resonance (FT-ICR). Each of these analyzers is discussed in this section, including their respective advantages and disadvantages. All mass analyzers distinguish ions according to their *m/z* ratio.

Quadrupole Analyzer

Quadrupoles are four precisely parallel rods with a direct current (DC) voltage and a superimposed radio-frequency (RF) potential (Figure 2.4). The field on the quadrupoles determines which ions are allowed to reach

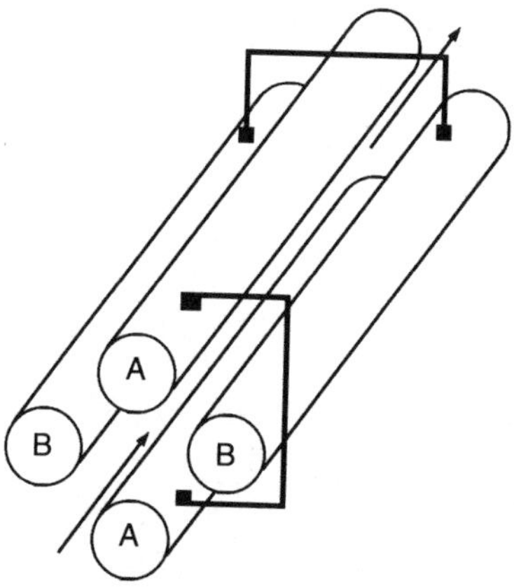

Figure 2.4 Longitudinal view of a quadrupole mass analyzer. The A rods are connected and are at the same DC and superimposed RF voltages. The same is true of the rods; however, they have an opposite DC voltage with respect to the A rods, and the RF field is phase shifted by 180°.

the detector. The quadrupoles thus function as a mass filter. As the field is imposed, ions moving into this field region will oscillate depending on their mass-to-charge ratio and, depending on the radio frequency field, only ions of a particular *m/z* can pass through the filter. The *m/z* of an ion is therefore determined by correlating the field applied to the quadrupoles with the ion reaching the detector. A mass spectrum can be obtained by scanning the RF field. Figure 2.5 illustrates the path of ions of different *m/z* values entering the quadrupole with a fixed DC and RF fields; only ions of a particular *m/z* are allowed to pass through.

Quadrupole mass analyzers have been used in conjunction with electron ionization sources since the 1950s. Electron ionization coupled with quadrupole mass analyzers are employed in the most common mass spectrometers in existence today. Quadrupole mass analyzers have found new utility in their capacity to interface with electrospray ionization. This interface has three primary advantages. First, quadrupoles are tolerant of relatively high pressures ($\sim 5 \times 10^{-5}$ Torr), which is well-suited to electrospray ionization since the ions are produced under atmospheric pressure conditions. Second, quadrupoles are now capable of routinely analyzing up to an *m/z* of 3000, which is useful because electrospray ionization of proteins and other bio-

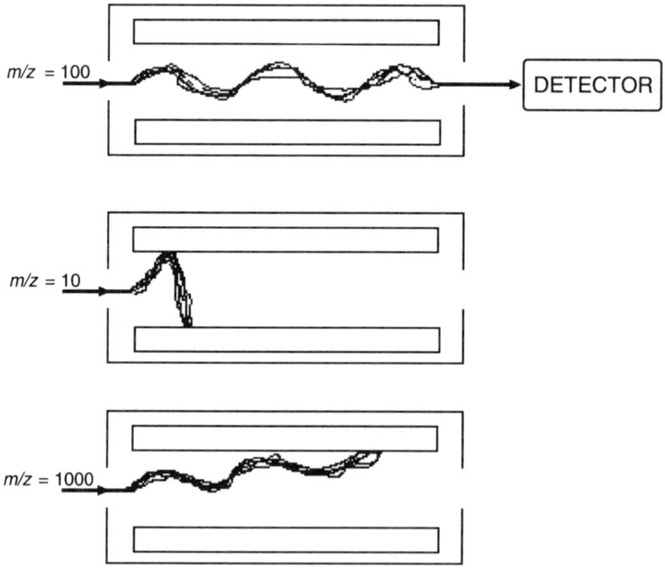

Figure 2.5 A cross section of a quadrupole mass analyzer taken as it analyzes for *m/z* = 100, 10, and 1000, respectively. It is important to note that both the DC and RF fields are the same in all three cases to allow only ions with *m/z* = 100 to traverse the total length of the quadrupole and reach the detector; the other ions are filtered out.

molecules commonly produces a charge distribution below m/z 3000. Finally, the relatively low cost of quadrupole mass spectrometers makes them attractive as electrospray analyzers. Considering these mutually beneficial features of electrospray and quadrupoles, it is not surprising that most of the successful commercial electrospray instruments thus far have been coupled with quadrupole mass analyzers.

Ion Trap

The ion trap mass analyzer shown in Figure 2.6 was developed at the same time as the quadrupole mass analyzer. The physics behind both of these analyzers is very similar; however, their applicability to electrospray ionization sources is still being developed. Their primary advantage is that tandem mass spectrometry experiments can be performed without having multiple analyzers.

Magnetic Mass Analyzer

As discussed in the beginning of the chapter, the earliest mass analyzers separated ions with a magnetic field. A moving charge passing through a magnetic field will experience a force, and travel in a circular motion with a radius of curvature depending upon the m/z of the ion. A magnetic analyzer separates ions according to their radii of curvature, and therefore only ions of a given m/z will be able to reach a point detector at any given magnetic field. A primary limitation of typical magnetic analyzers is their relatively low resolution.

Double-Focusing Magnetic Sector Mass Analyzer

In order to improve resolution, single-sector magnetic instruments have been replaced with double-sector instruments which combine the magnetic

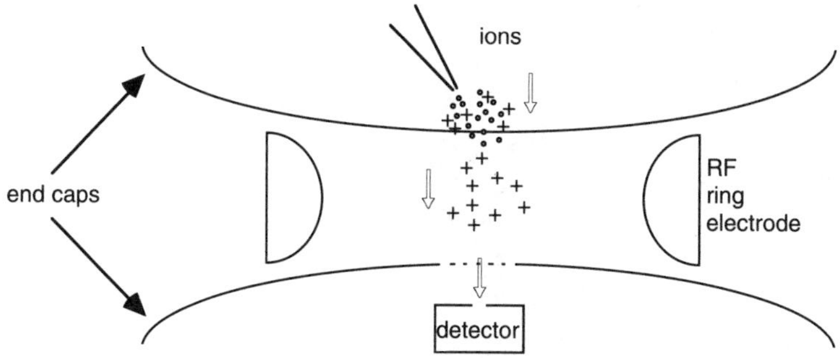

Figure 2.6 Ions inside an ion trap mass analyzer can be mass analyzed to produce a mass spectrum, or a particular ion can be trapped inside and made to undergo collisions to produce fragmentation information.

mass analyzer with an electrostatic analyzer. The two-sector instruments are called magnetic double-focusing mass analyzers, and can have resolving powers on the order of 100,000. The improved resolution obtained with a double-sector instrument is illustrated in Figures 2.7 and 2.8.

The magnetic double-focusing mass analyzer has two distinct parts, a magnetic sector and an electrostatic sector. The magnet serves the same purpose as described previously, to separate ions according to their mass-to-charge ratio; however, the ions will have a large kinetic energy distribution which limits the resolution attainable with only magnetic analysis (as shown in Figure 2.8).

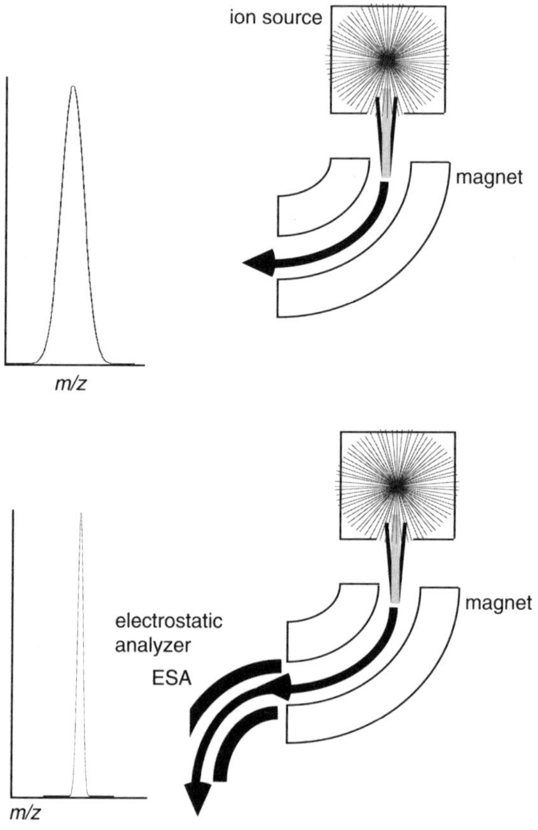

Figure 2.7 Illustration of the improved resolution obtained with a two-sector double-focusing instrument (below, magnet with electrostatic analyzer) over the single-focusing instrument (top, magnetic analyzer).

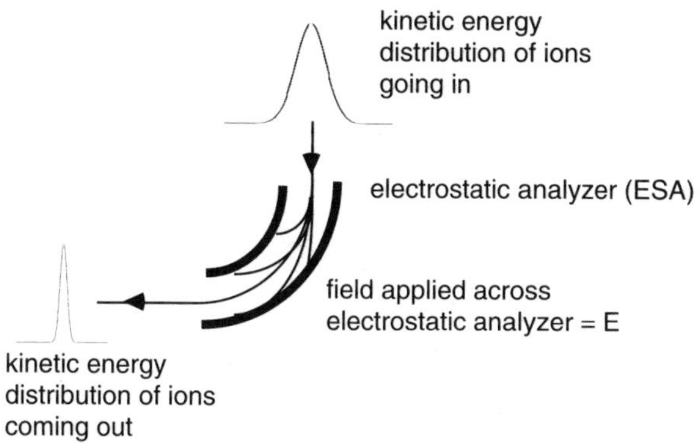

Figure 2.8 Illustration of the improved resolution obtained with the electrostatic analyzer when combined with magnetic separation of ions. The reduction in the kinetic energy spread directly relates to the spread of ions reaching the detector with respect to time. By reducing this distribution the overlap with other ions is reduced resulting in greater resolving power.

The electric sector acts as a kinetic energy filter allowing only ions of a particular kinetic energy to pass through its field, irrespective of their mass-to-charge ratio. Given a radius of curvature, R, and a field, E, applied between two curved plates (Figure 2.8), the equation $R = 2V/E$ allows one to determine that only ions of energy V will pass. Thus, the addition of an electric sector allows only ions of uniform kinetic energy to reach the detector, thereby increasing the resolution of the two sector instrument.

Magnetic double-focusing instrumentation is commonly used with FAB and EI ionization sources. Recently, this type of analyzer has also been used with electrospray and MALDI ionization sources.

Time-of-Flight Analyzer

A time-of-flight (TOF) analyzer is one of the simplest mass analyzing devices and is commonly used with MALDI ionization. Time-of-flight analysis is based on accelerating a set of ions to a detector with the same amount of energy. Because the ions have the same energy, yet a different mass (Figure 2.9), the ions reach the detector at different times. The process is analogous to a pitcher throwing a golf ball and a basketball at a catcher with the same amount of energy. The golf ball will reach the catcher faster because it has a smaller mass and therefore a greater velocity. So it is with ions. The smaller ions reach the detector first because of their greater velocity and the larger ions take longer, thus the analyzer is called time-of-flight because the m/z is determine from the ions' time of arrival.

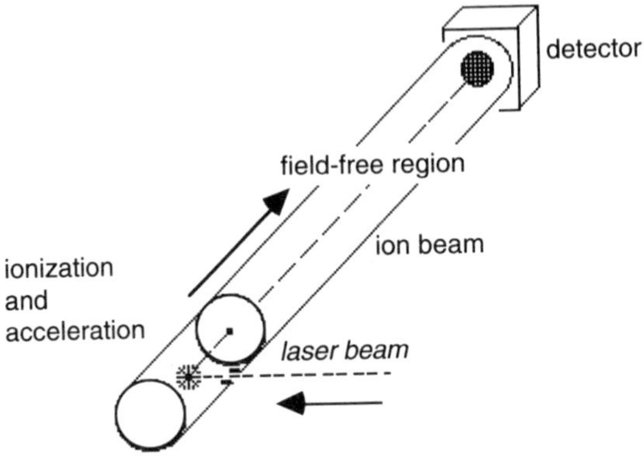

Figure 2.9 Time-of-flight mass analyzer.

The arrival time of an ion at the detector is dependent upon the mass, charge, and kinetic energy of the ion. Since kinetic energy (KE) is equal to $\frac{1}{2}mv^2$ or velocity $v = (2\mathrm{KE}/m)^{1/2}$, ions will travel a given distance, *d*, within a time, *t*, where *t* is dependent upon their *m/z*.

The resolving power of TOF instruments is low, but their main advantage is virtually no upper mass limitation.

Time-of-Flight Reflectron

Time-of-flight reflectron mass analysis (Figure 2.10) has been developed to help improve the resolution of time-of-flight analyzers, especially with MALDI. A time-of-flight analyzer with an attached reflectron is much like a magnetic sector with an electrostatic analyzer. While the time-of-flight analyzer has limited resolving power, the addition of the reflectron serves to reduce the kinetic energy distribution of ions that reach the detector and, as a result, achieve higher resolution. This increased resolution, however, comes at the expense of sensitivity and has a limited mass range.

Fourier Transform–Ion Cyclotron Resonance

ESI and MALDI–MS commonly use quadrupole and time-of-flight mass analyzers, respectively. The limited resolution offered by time-of-flight mass analyzers, combined with adduct formation observed with MALDI–MS, results in accuracy on the order of 0.1% to a high of 0.01%, while ESI typically has an accuracy on the order of 0.01%. Both ESI and MALDI are now being coupled to higher resolution mass analyzers such as the ultrahigh resolution ($>10^5$) mass analyzer. The result of increasing

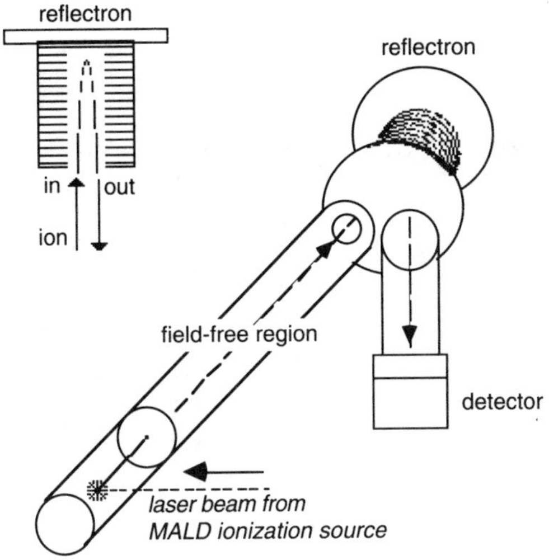

Figure 2.10 Time-of-flight reflectron analyzer improves resolution over the simple time-of-flight analyzer.

the resolving power of ESI and MALDI mass spectrometers is an increase in accuracy for biopolymer analysis.

Fourier transform–ion cyclotron resonance (FT-MS) offers two distinct advantages, high resolution and the ability to perform multiple collision events (MS^n, where n can be as high as 4). First introduced in 1974 by Comisarow and Marshall, FT-MS is based on the principle of a charged particle orbiting in the presence of a magnetic field (Figure 2.11). While the ions are orbiting, a RF signal is used to excite them and as a result of this RF excitation, the ions produce a detectable image current. The time-dependent image current can then be Fourier transformed to obtain the component frequencies of the different ions which correspond to their m/z.

Coupled to ESI and MALDI, FT-MS has potential to become an important research tool offering high accuracy with errors as low as $\pm 0.001\%$. The ability to distinguish individual isotopes of a protein of mass 29,000 is shown in Fig. 2.11.

The major characteristics of the mass analyzers are summarized in Table 2.1.

Tandem Mass Spectrometry

The new ionization techniques are relatively gentle and do not produce a significant amount of fragment ions, in contrast to EI which produces a

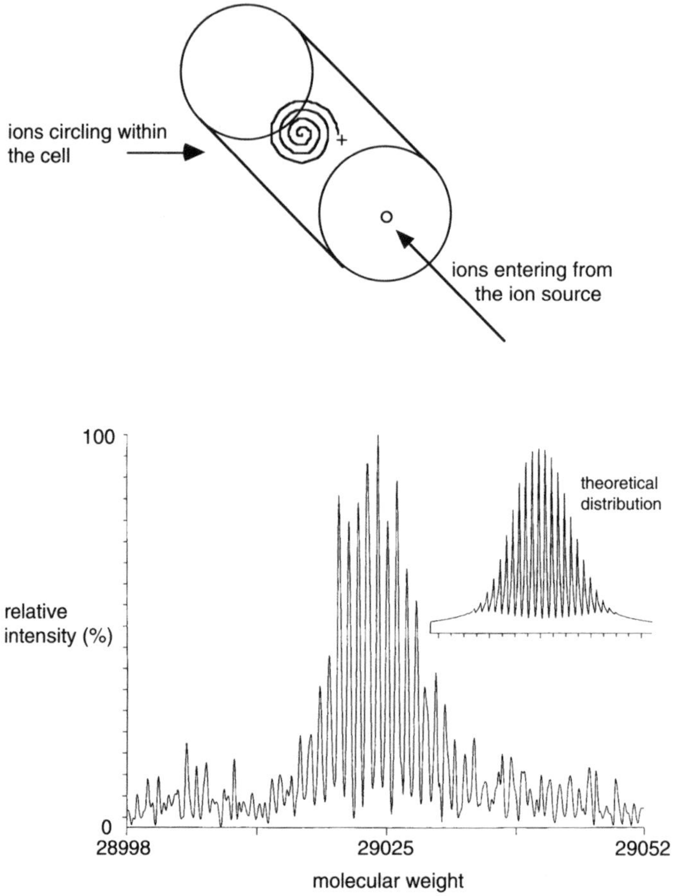

Figure 2.11 Fourier transform–ion cyclotron resonance data and mass analyzer. Single-scan ESI-generated data of carbonic anhydrase using a FT-MS analyzer. Adapted from Beu *et al.* (1993).

lot of fragment ions. Therefore, it has been necessary to develop techniques such as tandem mass spectrometry (MS/MS), to induce fragmentation. Tandem mass spectrometry (abbreviated MS^n—where n refers to the number of generations of fragment ions being analyzed) allows one to induce fragmentation and mass analyze the fragment ions. This is accomplished by collisionally generating fragments from a particular ion and then mass analyzing the fragment ions, as depicted in Figure 2.12. This section explores the technique, along with its application in conjuction with the previously discussed mass analyzers.

TABLE 2.1
General Comparison of Mass Analyzers

Mass analyzer	Typical mass range and resolution	Advantages	Disadvantages
Quadrupole	Range m/z 3000 Resolution 2000	Tolerant of high pressures Well-suited for electrospray Ease of switching between positive/negative ions Small size Relatively low cost	Mass range limited to about 3000 m/z Poor adaptability to MALDI
Ion trap	Range m/z 2000 Resolution 1500	Small size Medium resolution Simple design, low cost Well-suited for tandem mass spectrometry (MS^n, $n \leq 4$) Easy for positive/negative ions	Limited mass range of current commercial versions; however, progress is being made in their development
Magnetic sector	Range m/z 20,000 Resolution 10,000	Capable of high resolution Capable of exact mass Medium mass range Can be very reliable, manufacturer dependent	Not tolerant of high pressures Expensive Instrumentation is massive Relatively slow scanning
Time-of-flight (TOF)	Range $m/z\infty$ Resolution 350	Highest mass range Very fast scan speed Simple design, low cost Ease of adaptation to MALDI	Low resolution Difficulty of adaptation to electrospray
Time-of-flight reflectron	Range $m/z\infty$ Resolution 1500	Good resolution Very fast scan speed Simple design, low cost	Good resolving power has limited m/z range Lower sensitivity than TOF
Fourier transform–mass spectrometry (FT-MS)	Range m/z 10,000 Resolution 30,000	High resolution Well-suited for tandem mass spectrometry (MS^n, $n \leq 4$)	High vacuum ($<10^{-7}$ Torr) required Superconducting magnet required, expensive Instrumentation massive

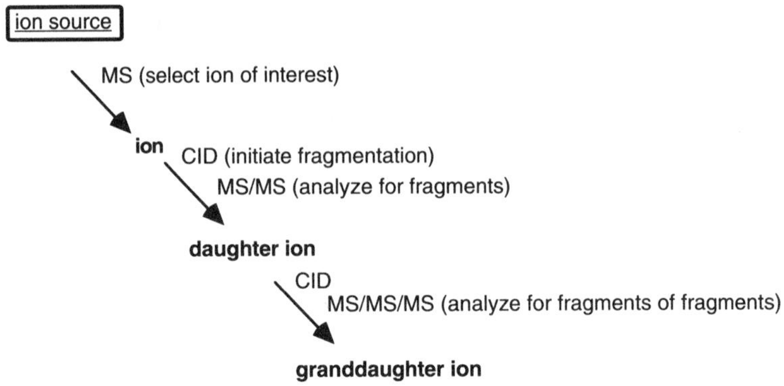

Figure 2.12 Tandem mass spectrometry, involving generation of fragment ions via collision-induced dissociation (CID) and the mass analysis (MS^n) of the progeny fragment ions. The terms parent, daughter, and granddaughter ions have been frequently used in the literature. Precursor, product, and second-generation product ions are also commonly used.

Fragmentation can be achieved by inducing ion–molecule collisions by a process known as collision-induced dissociation (CID) or collision-activated dissociation. CID is accomplished by selecting an ion of interest with a mass filter/analyzer and introducing that ion into a collision cell. A collision gas (typically Ar) is introduced into the collision cell, where the selected ion collides with the argon atoms, resulting in fragmentation. The fragments can then be analyzed to obtain a daughter ion spectrum. The abbreviation MS^n is applied to processes which analyze beyond daughter ions (MS^2) to granddaughter (MS^3), and to great-granddaughter ions (MS^4). Tandem mass analysis is primarily used to obtain structural information.

MS/MS with a Triple-Quadrupole Mass Spectrometer

For triple-quadrupole mass analyzers (Figure 2.13), tandem mass spectrometry utilizes the second quadrupole as a collision cell to generate fragment ions. The process of collision-induced dissociation is initiated by selecting an ion (e.g., $[\text{peptide} + H]^+$) having a particular m/z with the first quadrupole, Q1. Subsequently, only the selected ion (in this case $[\text{peptide} + H]^+$) is allowed to move into the second quadrupole (Q2; collision cell), where it colliides with argon atoms and fragments. These fragments can then be analyzed with the third quadrupole (Q3) and structural information can be obtained. This technique is used extensively in peptide and carbohydrate sequencing.

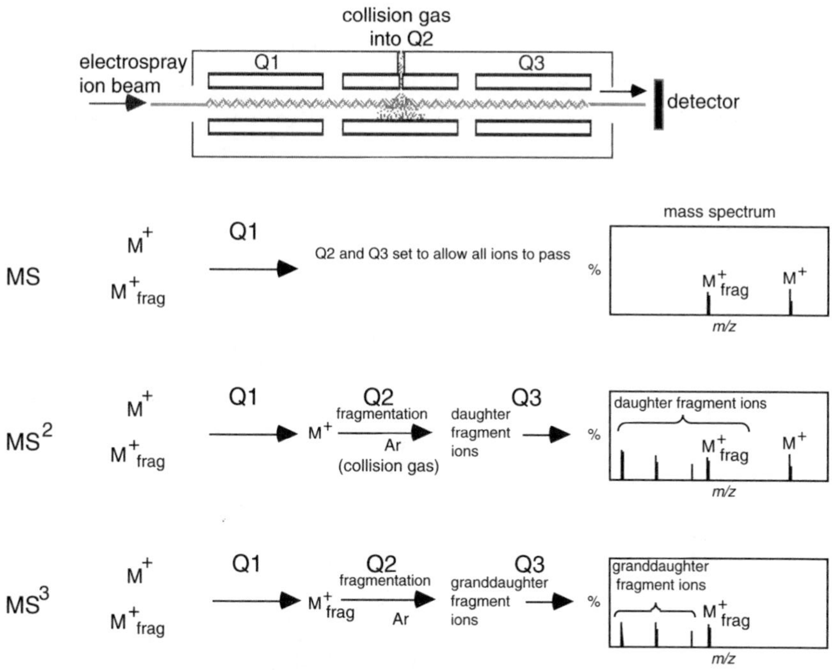

Figure 2.13 A triple-quadrupole ESI mass spectrometer possesses ion selection and fragmentation capabilities. Each quadrupole has a separate function: The first quadrupole (Q1) is used to scan across a preset *m/z* range or to select an ion of interest. The second quadrupole (Q2), also known as the collision cell, transmits the ions while introducing a collision gas (argon) into the flight path of the selected ion, and the third quadrupole (Q3) serves to analyze the fragment ions generated in the collision cell (Q2). For an MS experiment, Q1 scans over a selected *m/z* range and all the ions are observed. In the MS^2 experiment, the molecular ion M^+ can be selected by Q1, which results in its fragmentation at Q2. Analysis of the fragments occurs at Q3. An MS^3 experiment can be performed if a daughter fragment ion is generated in electrospray ionization, i.e., M^+_{frag}. Q1 can be used to select the daughter ion, Q2 will generate granddaughter fragment ions, and Q3 will mass analyze for the granddaughter fragment ions.

MS/MS with a Two-Sector Mass Spectrometer (Linked Scanning)

Two-sector or double-focusing mass spectrometers were originally developed to perform accurate mass measurements. However, it became apparent that they could be scanned in special ways so that metastable decompositions could be observed without interference from source-generated ions. This was accomplished by linking the magnetic and electrostatic fields such that they were scanned together. The result of this linking allows the researcher to obtain reasonable product ion spectra; however, precursor ion selection is poor. These instruments are not considered tandem mass spectrometers, yet they merit mention because of the tandem-like information they can give.

MS/MS with a Four-Sector Mass Spectrometer (Magnet/ESA–Magnet/ESA)

MS/MS can also be performed with magnetic sector instruments in tandem, as shown in Figure 2.14. The ions can be selected out by the first two sectors, followed by mass analysis in the second two sectors. As with the triple quadrupole, structural information can be obtained. Four-sector MS/MS is used in peptide and carbohydrate sequencing, its primary advantage being that high-energy collisions cause significant fragmentation and therefore facilitate the acquisition of detailed structural information. The primary disadvantage of four-sector MS/MS is its cost and size.

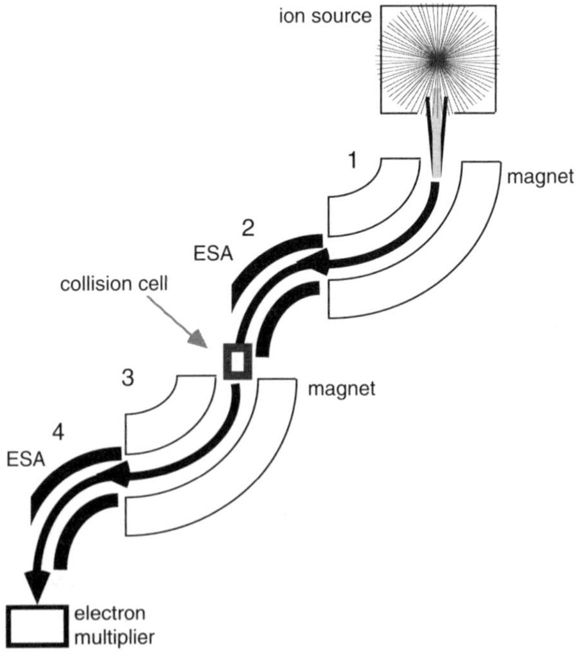

Figure 2.14 Four-sector magnetic/electrostatic mass analyzer used for collisional studies.

MS/MS with a Two-Sector (Magnetic/ESA) Quadrupole Mass Spectrometer

MS/MS with a two-sector magnetic/electrostatic mass analyzer coupled to a quadrupole mass analyzer is also common (see Figure 2.15). These hybrid instruments, as they are called, are not as sensitive or as accurate as the four-sector magnetic/electrostatic instruments.

MS/MS with a Time-of-Flight Reflectron Mass Spectrometer

Performing MS/MS experiments is also possible with time-of-flight reflectron mass spectrometers. This application has just begun to be used with MALDI and shows promise in obtaining sequencing information on biomolecules. MS/MS is accomplished by taking advantage of fragmentation that occurs following ionization or postsource decay. Time-of-flight instruments alone will not separate postionization fragment ions because the fragment ions all have the same velocity and thus reach the detector at the same time as the precursor ion. The reflectron takes advantage of the fact that the fragment ions have different kinetic energies than their precursor ion. The reflectron separates the fragment ions with respect to kinetic energy and thus produces a fragment ion spectrum (Figure 2.16).

MS/MS with a Fourier Transform–Ion Cyclotron Resonance

In addition to its high sensitivity, Fourier transform–ion cyclotron resonance is very useful for multiple collision experiments (MS^n, $n \leq 4$).

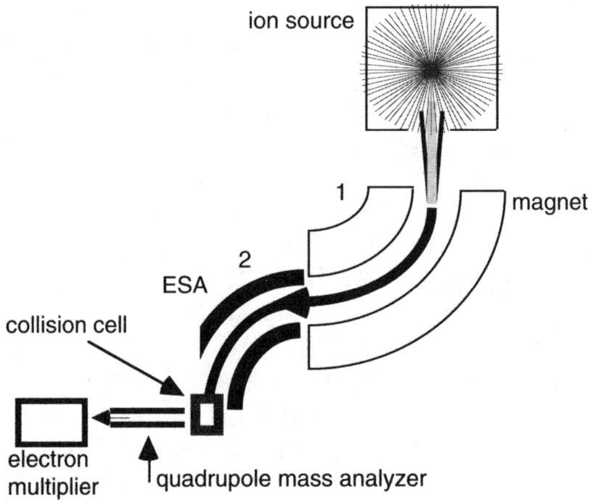

Figure 2.15 Illustration of a two-sector instrument with detection of fragment ions using a quadrupole mass analyzer.

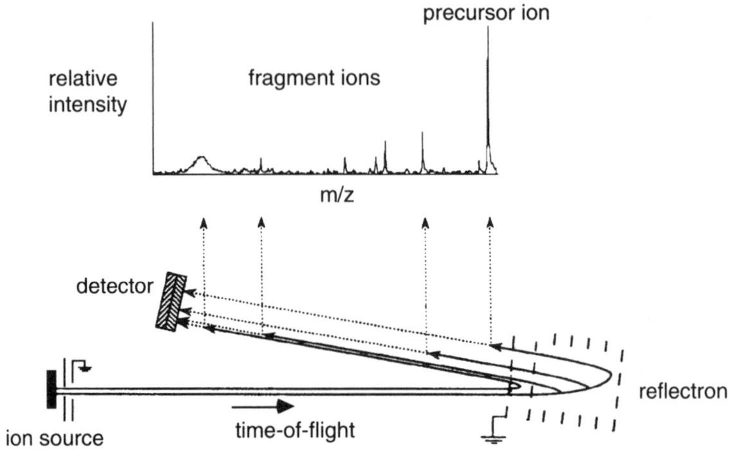

Figure 2.16 Illustration of a time-of-flight reflection mass analyzer and its ability to generate fragmentation information as shown in the mass spectra (top). Adapted from Verentchikov *et al.* (1994).

FT-MS is capable of ejecting all but ions of a selected *m/z* from the cell, after which a collision gas can be introduced into the cell to induce fragmentation. Mass analysis can then be carried out on a fragment or a fragment of a fragment. Multiple collision experiments with FT-MS, followed by mass analysis, are a unique way of accomplishing tandem mass spectrometry. The high resolution of FT-MS/MS yields high-accuracy of the fragment ions.

MS/MS with an Ion Trap

The ion trap is also useful for tandem mass measurements, providing structural information on biomolecules, and is just beginning to be offered as a mass analyzer with electrospray ionization.

The characteristics of tandem mass analyzers are summarized in Table 2.2.

Ion Detectors

Passing through the mass analyzer, we now come to the final element of the mass spectrometer, the ion detector (Figure 2.17). The detector allows a mass spectrometer to generate a signal (current) from incident ions, by generating secondary electrons, which are further amplified, or by inducing a current generated by a moving charge. Among the detectors described below, the electron multiplier and scintillation counter are proba-

TABLE 2.2
General Comparison of Tandem Mass Analyzers

Mass analyzer	Advantages	Disadvantages
Tandem quadrupoles	Good accuracy/resolution	Low-energy collisions resulting in incomplete fragmentation
Two sector	Much less expensive than four sector	Low sensitivity Low resolution/accuracy
Four sector	Resolution/accuracy high High-energy collisions—more complete sequencing information	Equipment is massive Expensive
Two-sector with quadrupole	High energy collisions—more complete sequencing information Resolution/accuracy good	Reasonable sensitivity Medium cost
Time-of-flight reflection	Relatively inexpensive	Precursor ion selection is somewhat limited to a wide mass range Low resolution of product ions
Fourier transform–Ion cyclotron resonance (FT-MS)	High resolution of product ions Well suited for MS^n ($n \leq 4$)	Low-energy collisions Requires a high vacuum and a superconducting magnet
Ion trap	Well suited for MS^n ($n \leq 4$) Relatively inexpensive Good accuracy/resolution	Low-energy collisions

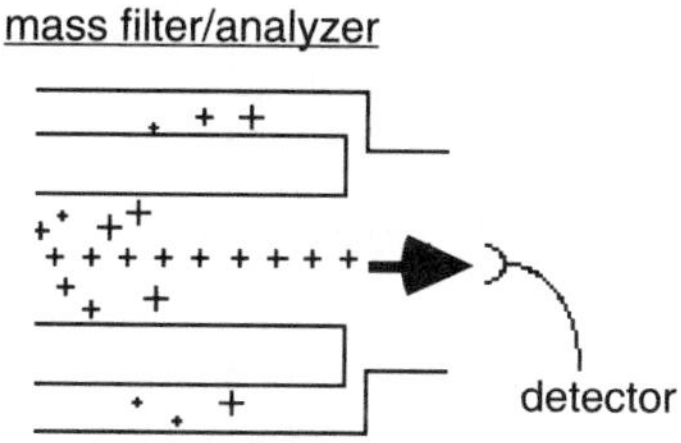

Figure 2.17 Ion detection.

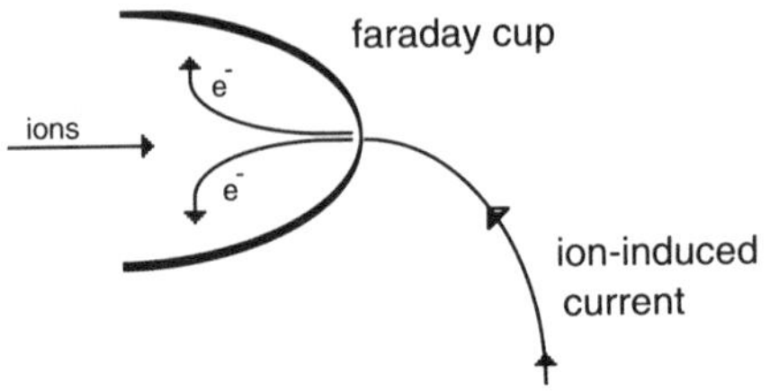

Figure 2.18 Faraday cup converts the striking ion into a current.

bly the most commonly used, converting the kinetic energy of incident ions into secondary electrons.

Faraday Cup

A Faraday cup (Figure 2.18) operates on the basic principle that a change in charge on a metal plate results in a flow of electrons and therefore creates a current. One ion striking the dynode surface of the Faraday cup (a dynode is a secondary emitting material, usually BeO, GaP, or CsSb) induces several secondary electrons to be ejected and temporarily displaced. This temporary emission of electrons induces a current in the cup and provides for a small amplification of signal when an ion strikes the cup. This detector is relatively insensitive, yet robust and simple in design.

Electron Multiplier

An electron multiplier (Figure 2.19) is one of the most common means of detecting ions, achieving high sensitivity by extending the principle used with a Faraday cup. Whereas a Faraday cup uses one dynode, an electron multiplier is made up of a series of dynodes maintained at ever-increasing potentials. Ions strike the dynode surface, resulting in the emission of electrons. These secondary electrons are then attracted to the next dynode where more secondary electrons are generated, ultimately resulting in a cascade of electrons. Typical amplification or current gain of an electron multiplier is 10^6, with a lifetime of 1 to 2 years. The lifetime of the electron multiplier is limited because of surface contamination from incident ions or from a relatively poor vacuum.

Photomultiplier Conversion Dynode (Scintillation Counting or Daly Detector)

The photomultiplier conversion dynode detector in Figure 2.20 is similar to an electron multiplier where the ions initially strike a dynode, resulting in the emission of electrons. However, with the photomultiplier conversion

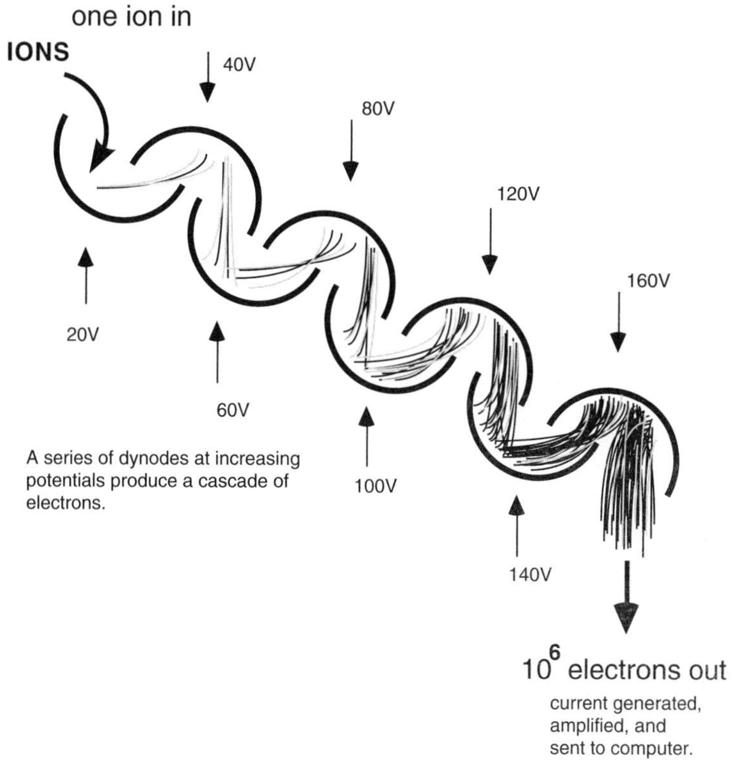

Figure 2.19 Electron multiplier and the cascade of electrons that results in a 10^6 amplification of current in a mass spectrometer.

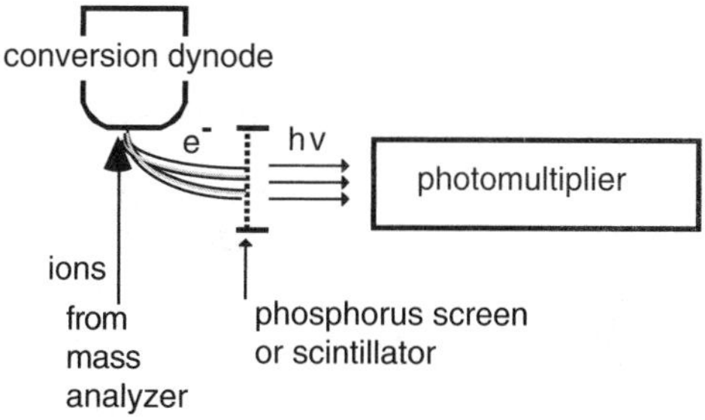

Figure 2.20 Scintillation counting relies on the conversion of the ion/electron signal into light. Once the photon(s) are formed, detection is performed with a photomultiplier.

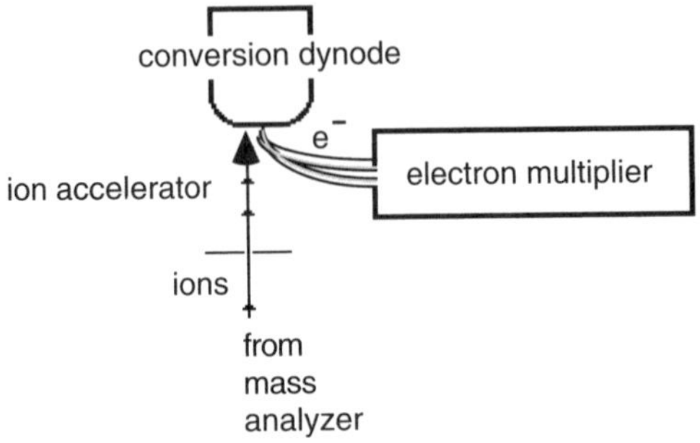

Figure 2.21 High-energy dynode detector.

dynode detector electrons then strike a phosphorus screen. The phosphorus screen, much like the screen on a television set, releases photons once an electron strikes. These photons are then detected by a photomultiplier, which operates with a cascading action much like an electron multiplier. The primary advantage of the conversion dynode setup is that the photomultiplier tube is sealed in a vacuum (photons pass through sealed glass), unexposed to the internal environment of the mass spectrometer. Thus the possibility of contamination is removed. A 5-year or greater lifetime is typical and, with sensitivity similar to that of electron multipliers, photomultiplier conversion dynode detectors are becoming more widely used in mass spectrometers.

High-Energy Dynode Detector (HED)

The HED (Figure 2.21) uses an electrostatic field prior to the electron multiplier. Once an ion enters the field, it is accelerated into the electron multiplier, thus producing more secondary electrons which further cascade

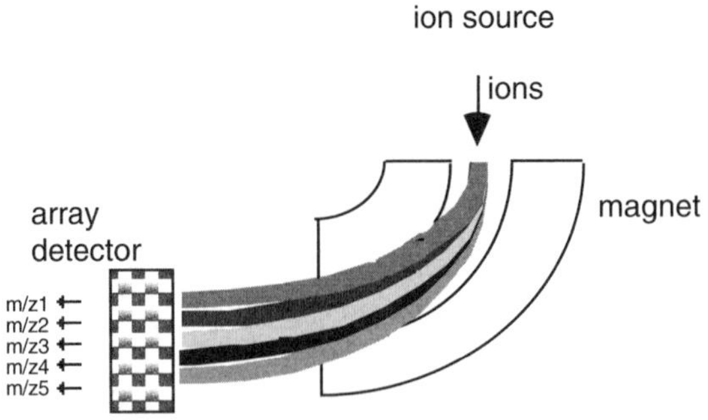

Figure 2.22 Illustration of the operation of an array detector.

in the electron multiplier. HEDs serve to increase ion energy and therefore the signal intensity, which results in greater sensitivity.

Array Detector

An array detector (Figure 2.22) is essentially a group of detectors in a linear array. This array allows for a group of ions (having different *m/z*) to be detected simultaneously after passing through a magnetic sector analyzer. The *m/z* of an ion is determined by where it strikes the detector. Since scanning is not necessary, array detection is most useful for high-sensitivity work; however, it can only be used over a small mass range.

Fourier Transform–Ion Cyclotron Resonance Mass Spectrometer

The FT–MS mass analyzer also functions as the detector because the moving ions induce a current corresponding to their *m/z*.

The characteristics of ion detectors are summarized in Table 2.3.

TABLE 2.3
General Comparison of Ion Detectors

Detector	Advantages	Disadvantages
Faraday cup	Good for checking ion transmission and detector sensitivity	Low amplification (≈10)
Scintillation counter	Extremely robust Long lifetime (>5 years) Good sensitivity (≈106)	Cannot be exposed to light
Electron multiplier	Fast response Sensitive (≈106)	Short lifetime (1–2 years)
High-energy dynodes with electron multiplier	Increases high mass sensitivity	May shorten lifetime of electron multiplier
Array	Fast and sensitive	Low resolution ~0.2 Da Expensive Short lifetime (<1 year)
FT-MS	Mass analyzer is the detector High resolution (~500,000)	Used only with FT-MS instruments

Summary

Analyzers

Mass analyzers scan or select ions over a particular *m*/*z* range. The key feature of these analyzers is that they measure the *m*/*z* ratio, *not* the mass.

Quadrupoles function as a mass filter, allowing only ions of a particular *m*/*z* to pass through. As the quadrupoles are scanned, an entire mass range of ions is allowed to pass through during the scan period.

A two-sector instrument has two distinct parts, a magnetic analyzer and an electrostatic analyzer. The magnet serves to separate ions with limited resolving power, and the electrostatic analyzer filters the ions according to their kinetic energy, irrespective of their mass-to-charge ratio. The energy filtering allows for high resolution, and is commonly used for exact mass analysis required in the identification of synthetic organic compounds.

Time-of-flight mass analysis is the result of accelerating ions of differing *m*/*z* through the same kinetic energy. Ions (directed into a field-free region) will reach the detector at different times depending on their *m*/*z*. A time-of-flight analyzer is one of the simplest analyzing devices.

A time-of-flight reflectron analyzer is much like a magnetic sector with an electrostatic analyzer. The time-of-flight analyzer is capable of achieving limited mass resolution, while the addition of the reflectron serves to reduce the kinetic energy distribution. As a result, higher resolution is achieved with the reflectron. However, this increased resolution comes at the expense of sensitivity. TOF reflectron instruments maintain this increased resolution only for a limited mass range.

FT–MS and ion traps are commercially newer analyzers and have the potential to play a major role in the future of mass spectrometry.

Tandem Mass Spectrometry

Tandem mass spectrometry allows for the production and identification of fragment ions from a selected ion. Fragmentation can be achieved by inducing ion–molecule collisions by a process known as CID. CID is accomplished by selecting an ion of interest with a mass filter/analyzer, followed by the introduction of that ion to a collision gas, typically argon. The spectral information obtained from tandem mass analysis is primarily structural information.

Detectors

Ions are detected by converting their kinetic energy into an electrical current. Among the four most common types of detectors, the electron multiplier and scintillation counter are the most frequently used. An electron multiplier is made up of a series of approximately 10 dynodes, each made of a secondary emitting material (usually BeO, GaP, or CsSb) maintained at ever-increasing potentials. Ions strike the dynode surface, which results in the emission of electrons. These secondary electrons are then attracted to the next dynode and then the next, resulting in a cascade. Typical current gain is about 10^6. The scintillation counter (photomultiplier conversion dynode) detector is similar to an electron multiplier. The ions initially strike a dynode, thus releasing electrons; however, in this case the electrons then strike a phosphorous screen. When the phosphorous screen is struck by an electron it releases photons, which are then detected by a photomultiplier.

Practice Questions

1. What does a mass analyzer measure?
2. What makes magnetic sector instruments useful for fragmentation studies?
3. What makes magnetic sector instruments prohibitive for fragmentation studies?
4. Why wouldn't time-of-flight analysis be useful for high-accuracy work?
5. Why are conversion dynode detectors so durable?
6. What does MS^n ($n \leq 4$) mean?
7. Which analysis technique shows the most promise for the future? Why?

Answers to Practice Questions

1. The mass-to-charge ratio.
2. High-energy collisions.
3. The size and correspondingly high cost.
4. Time-of-flight analysis has low resolution.
5. The photomultiplier is kept sealed in its own vacuum.
6. MS^n refers to the number of MS experiments performed. MS^3 means that a parent ion was selected (MS^1) and exposed to a collision gas, resulting in the formation of fragment ions (MS^2). One of the resulting fragment ions was selected and further exposed to collision gas, and the resultant granddaughter ions were detected (MS^3).
7. FT-MS probably shows the most promise. FT-MS offers very high accuracy as a straight mass analyzer and as a tandem mass analyzer as well.

References

Beu, S. C., Senko, M. W., Quinn, J. P., and McLafferty, F. W. (1993). *J. Am. Soc. Mass Spectrom.* **4,** 190–192.

Verentchikov, A. N., Ens, W., and Stangind, K. G. (1994). *Anal. Chem.* **66,** 126–133.

CHAPTER 3

Basics of Biomolecule Mass Spectrometry

This chapter examines the general application of mass spectrometry to biomolecules. As compounds are being developed in the hope of creating new treatments for cancer, AIDS, and other diseases, mass analysis offers a routine and relatively easy means of characterization. For such routine mass analysis to be successful, a few simple conditions must be met. This chapter familiarizes the reader with these general conditions as well as sample preparation.

General Preparative Techniques (FFAST–MS)

In the 1960s and 1970s, biomolecule mass analysis was typically accomplished by adding protecting groups to make molecules more volatile and stable. The stable molecule could then be thermally vaporized into the mass spectrometer's ionization source, which utilized either electron or chemical ionization. Even with derivatization, this type of mass analysis was limited to molecular weights of 1000 Da. Also, derivatization techniques were laborious and often required large amounts of sample. This changed in the 1980s with the development of the FAB, ESI, and MALDI ionization sources described in the previous two chapters.

Barbara Larsen (Dupont) recommends some of the following elements to help obtain good mass data [Charles McEwen and Barbara Larsen (1992), A Course on Peptide and Protein Analysis, *American Society for Mass Spectrometry*, Washington, DC]. Listed here according to the acronym FFAST–MS, these recommendations hold equally well for FAB, MALDI, and electrospray:

F F A S T - M **S**alt content should be miniminized
F F A S T - **M**atrix selection/preparation

F F A S **T**idy, care should be taken to maintain high purity
F F A **S**olubility, sample solubility in the solvent or matrix is crucial
F F **A**mount, is there enough sample?
F **F**unctional groups, the functional groups help determine how to analyze
Fast, analyze soon after synthesis and/or purification

Fast

By analyzing the sample soon after it has been isolated/purified you reduce the possibility of degradation.

Functional Groups

The types of functional groups on a molecule will often determine how a compound should be analyzed. When analyzing a compound by FAB, ESI, or MALDI, first look for sites on the molecule that can be ionized. Amines (+), acids (−), and amides (+) represent easily ionizable functional groups, while hydroxyl groups, esters, ketones, and aldehydes are sites that do not accept a charge as easily and are therefore more difficult to ionize, typically resulting in weak ion signals. Due to the presence of amide and amine groups, peptides usually ionize easily through protonation. Because peptides often contain acids, they can also be observed in the negative ion mode by deprotonation. Some carbohydrates will accept a proton because of the presence of an amide bond, but in general they form stable ions upon the addition of a cation other than a proton, such as Na^+ or K^+. These cations can be added to form $M + Na^+$ and $M + K^+$ ions. Oligonucleotides, proteins, and the myriad of small molecules each have their ionization peculiarities, making the conditions under which analysis is performed all the more important (Scheme 3.1). Table 3.1 describes some of the different ways different compounds are analyzed.

Amount

The amount of sample is critical. Too little or even too much sample can dramatically affect the mass spectrum and mean the difference between a successful and an unsuccessful analysis. If too little of a sample is used, the instrument will be unable to detect. Too much sample can saturate the detector and skew the mass by causing ^{13}C peaks to dominate or make impurities appear more dominant. Analyzing too much sample can even cause signal suppression. The point is that more sample is not always better and it is important to be in the right range.

The amount of sample required varies according to the technique used. Table 3.2 converts a range of concentrations often used in electrospray and MALDI from micromolar (20–50 μM) to milligrams/milliliter with respect

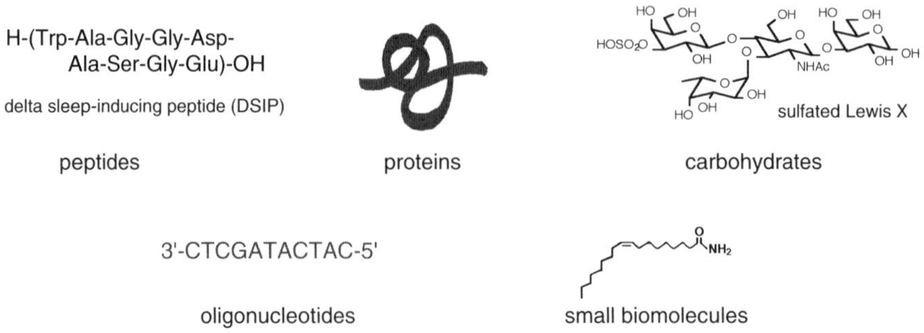

Scheme 3.1

TABLE 3.1
Typical Ionization Properties and Techniques for Biomolecules

Compound	Ionization mechanism	Ionization techniques	Ionization mode
Peptides	Protonation	FAB, MALDI, and ESI	Positive
	Deprotonation	FAB, MALDI, and ESI	Negative
Proteins	Protonation	MALDI and ESI	Positive
Membrane proteins	Protonation	MALDI and ESI	Positive
Glycoproteins	Protonation	MALDI and ESI	Positive
Carbohydrates	Protonation	FAB, MALDI, and ESI	Positive
	Cationization	FAB, MALDI, and ESI	Positive
	Deprotonation	FAB, MALDI, and ESI	Negative
Protected carbohydrates	Protonation	FAB, MALDI, and ESI	Positive
	Cationization	FAB, MALDI, and ESI	Positive
	Deprotonation	FAB, MALDI, and ESI	Negative
Oligonucleotides	Deprotonation	MALDI and ESI	Negative
Protected oligonucleotides	Deprotonation	MALDI and ESI	Negative
	Protonation	MALDI and ESI	Positive
	Cationization	MALDI and ESI	Positive
Small biomolecules	Protonation	FAB, MALDI, and ESI	Positive
	Cationization	FAB, MALDI, and ESI	Positive
	Deprotonation	FAB, MALDI, and ESI	Negative
	Electron ejection	Electron ionization	Positive
	Electron capture	Electron ionization	Negative

TABLE 3.2
Conversion for the Relevant Concentrations in Electrospray, MALDI, and FAB (micromolar to milligrams per milliliter, μM to mg/ml) with Respect to Molecular Weight[a]

Molecular weight (Da)	Electrospray or MALDI (20–50 μM)	FAB (≥ 1.0 mM)
500	0.01–0.025 mg/ml	≥0.50 mg/ml
1,000	0.02–0.05 mg/ml	≥1.00 mg/ml
2,500	0.05–0.13 mg/ml	≥2.50 mg/ml
5,000	0.10–0.25 mg/ml	≥5.00 mg/ml
10,000	0.20–0.50 mg/ml	≥10.00 mg/ml
20,000	0.40–1.00 mg/ml	
30,000	0.60–1.50 mg/ml	
40,000	0.80–2.00 mg/ml	
50,000	1.00–2.50 mg/ml	
60,000	1.20–3.00 mg/ml	
70,000	1.40–3.50 mg/ml	
80,000	1.60–4.00 mg/ml	
90,000	1.80–4.50 mg/ml	
100,000	2.00–5.00 mg/ml	

[a] Typical sample volumes used for ESI, MALDI, and FAB are ≤50, ≤5, and ≤5 μl, respectively.

to molecular weight. Those used for FAB analysis are also displayed. The concentrations shown are conservative; most instruments will easily handle significantly lower concentrations. In general, however, the concentration listed will produce a good signal even in the presence of some salt or other contaminating compounds.

In comparing electrospray, MALDI, and FAB, Table 3.2 illustrates the relative insensitivity of FAB. This lack of sensitivity can be a surprisingly useful feature, as FAB tolerates large quantities of sample (from 1 μg to 1 mg), and a thorough cleaning (10 sec of wiping the probe tip with a methanol-soaked cotton swab) will ensure virtually no carryover. This utility is especially apparent with synthetically derived compounds, where there is usually an abundance of sample for mass analysis. Electrospray is more sensitive, necessitating constant monitoring of the instrument from sample to sample to ensure that the observed ions are not carryover from a previous sample. This is imperative if too much sample was used before, or if the instrument is particularly sensitive to a previously analyzed compound, MALDI is useful in this respect because the probes, similar to the FAB probes, are relatively easy to clean or are disposable, and sensitivity is not sacrificed with MALDI.

Using an excessive amount of sample in electrospray can have dire consequences. One result will be signal saturation. If this occurs, the mass of the ion of interest may be skewed because of detector saturation of the ^{12}C isotope and an apparent increase in the contribution from the ^{13}C isotope (Figure 3.1B). Saturation will also produce excessive ion formation of relatively low quantities of impurities (Figure 3.1B). As a result the sample can show up in future spectra and require numerous rinses to remove (in some cases it takes hours to remove excessive sample from a contaminated instrument). One must accept the problems associated with contamination and saturation, as they are outweighed by the excellent sensitivity offered by electrospray.

Solubility

How well a sample dissolves in the solvent or matrix solution is critical to obtaining mass data. The solvent or matrix is the medium by which your compound will be transported to the gas phase and it often provides the conditions that make ionization possible. Whether performing electrospray, MALDI, or FAB, if solubility is poor, it will be difficult or simply not possible to vaporize or ionize the sample molecule. If the compound is not

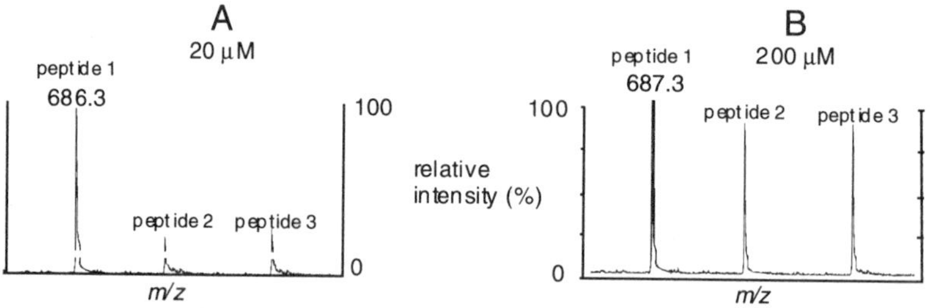

Figure 3.1 The problem of detector saturation is shown in these electrospray mass spectra of a mixture of peptides. Spectrum **A** contains *peptide 1* at a concentration 20 *μM,* while spectrum **B** was taken at a 10-fold higher concentration (200 *μM*). Relatively minor impurities, *peptide 2* and *peptide 3,* observed in spectrum **A** as relatively weak signals, now appear as major components in spectrum **B.** In addition, the signal of *peptide 1* has been so saturated that the ^{13}C of *peptide 1* dominates in spectrum **B** and the mass of the peptide appears to be bigger. These spectra demonstrate the problem of analyzing a sample at too high a concentration.

soluble in the solvent or matrix, it will be virtually impossible to obtain any signal at all. Figure 3.2 illustrates ESI data obtained on a protein in unsuitable and suitable solvent systems.

When submitting a sample for mass analysis, it is very important to provide solubility information. A simple note describing what solvent is most suitable can be invaluable. Our laboratory uses a "Mass Spectrum Request and Analysis Form," which communicates the important information needed to obtain mass data simply through circled items on the form (Figure 3.3). The upper portion is completed by the investigator requesting analysis. It includes information such as structure, formula, and confidence in stated information. The bottom portion is the analysis, which gives the results, ionization tool used, solvent and/or matrix used, and the resolution of the instrument—basically, the general conditions under which the analysis was performed.

Purity

The more pure or homogenous, the better chance of obtaining an accurate and clean mass spectrum. Electrospray requires that the sample be relatively free of salts and detergents, with contamination less than millimolar. While MALDI is known for giving good signals even in the presence of salt and impurities, the electrospray signal is more susceptible to these contaminants and will result in the loss of sensitivity. The loss of

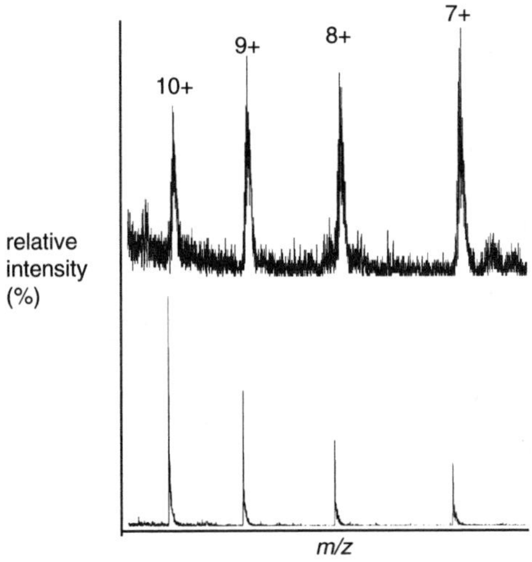

Figure 3.2 A 14-kDa protein electrospray mass analyzed in methanol (top) and in 33% methanol, 33% acetonitrile, and 33% water (bottom).

Mass Spectrum Request and Analysis Form

The Scripps Research Institute Mass Spectrometry Laboratory (619) 554-9415

2 Digit Sample I.D. #	Molecular Formula	Monoisotopic Mass
99	$C_{30}H_{19}N_1O_9$	*537.1*

Structure

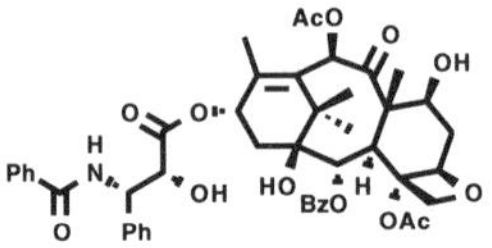

Precision Required:	[Exact and Unit Mass]		Unit Mass
Confidence in Structure:	Tentative	[Confident]	**Confirmed**
Purity:	Crude	[Relatively Pure]	**Very Pure**
Toxicity:	[Extremely Toxic]	Toxic	**Safe**

Are any buffers or alkali cations present, e.g., Li^+, Na^+, K^+, Rb^+, and/or Cs^+? *YES, Na^+*

FAB/EI solution concentration ___*4.0*___ mg/ml (~5.0 mg/ml), solvent is *MeOH*____
ES/MALDI concentration __________ μM (~20-50 μM), solvent is __________

Suitable Solvents? [MeOH] $CHCl_3$ H_2O [DMSO] THF Other________

Analysis - **Analysis**

Comments		MW(expected)		MW(observed)	error	
EI+ or [FAB+]/-					2.7	mmu
MALD+/-	**M + Cs^+**	**670.0114**	exact	**670.0141**	4.0	ppm
Electrospray+/-		**670**	unit	**670**		
GC/MS 50° $\xrightarrow{20°/min}$ 300°						

Solvent Used:	[MeOH]	$CHCl_3$	CH_2Cl_2	TFA	______	**Resolution:**	
Matrix Used:	NBA	DHB	[NBA/CsI]	Gly	______	1000 [5000]	________

Figure 3.3 Example of a mass spectrum request form. Taxol was synthesized by K. C. Nicolaou and his group at The Scripps Research Institute. The bottom portion was completed by the operator. DMSO, THF, and TFA are abbreviations for dimethylsulfoxide, tetrahydrofuran, and trifluoroacetic acid, respectively.

sensitivity associated with heterogeneous electrospray samples is possibly due to competition for cations in the droplets, and/or the effect impurities may have on droplet desolvation. If a droplet cannot effectively evaporate because of reduced vapor pressure associated with impurities, you will lose or stop ion production.

To avoid inhibiting ion production and to increase sensitivity, purification procedures such as reverse-phase liquid chromatography (protein, peptides, and DNA), normal-phase chromatography, and dialysis are commonly used.

These traditional cleanup procedures are also useful for MALDI samples. MALDI has also seen a dramatic increase in sensitivity as a result of some recently developed sample preparation techniques. These techniques allow for the acquisition of a uniform signal across the matrix (some of these techniques are described in Chapter 4). Typically, with MALDI you will hunt for a good signal by moving the laser beam around the probe until a reasonable signal has been found. A good signal is the result of the laser beam striking a portion of the probe that has the sample embedded in crystalline matrix. These areas are also known as hot spots on the probe's surface. With a newer technique, Professor Ron Beavis has emphasized the importance of uniformly depositing the matrix material onto the surface of the MALDI probe. Once a thin uniform coating of matrix material has been deposited onto the probe tip, the sample can be introduced and finally washed clean. This method creates a probe surface where uniform ion desorption is observed. In other words, significant desorption is achieved across the whole surface of the proble, making it less necessary to hunt for the signal. The following section on matrix selection/preparation further describes these cleaning procedures.

Matrix Selection/Preparation

The choice of matrix is critical in mass analysis and, in many cases, will determine the quality of the data. Sample preparation for FAB analysis is a simple procedure that works very well for many samples. It is accomplished by dissolving the sample in a cosolvent at a concentration of approximately 1–5 mg/ml and then adding approximately 2–4 μl of this solution to the matrix (typically nitrobenzyl alcohol, shown below). The sample solution is dissolved in the matrix on a metal probe tip, which is generally a flat round surface approximately 2–5 mm in diameter. Cosolvents that work well with the NBA matrix are 5–50% acetic acid in water, acetonitrile, chloroform, and methanol. Formation of a protonated molecular ion can also be facilitated by the addition of a small amount of TFA (0.1–25%). Some of the more common matrix materials for FAB are shown in Figure 3.4.

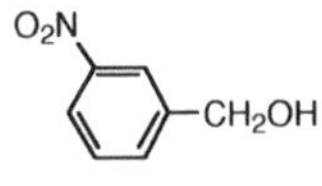

m-nitrobenzyl alcohol
(NBA)
a standard matrix in many laboratories

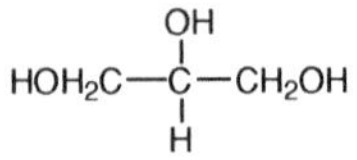

glycerol

original FAB matrix material
still widely used

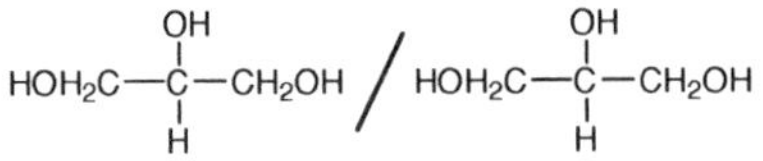

glycerol/thioglcerol
(Gly/TGly)
utility similar to glycerol, may work
better for large and highly ionic compounds,
has an odor

$N(CH_2CH_2OH)_3$

triethanolamine
(TEA)
used primarily in negative ion analysis

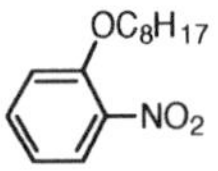

2-nitrophenyl octyl ether
(NPOE)
aprotic, useful for reactive compounds

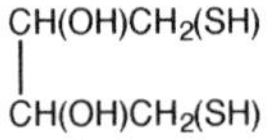

(isomers)

1:5 dithiothrietol/dierythritol
(magic bullet)
also a useful matrix, has an odor

Figure 3.4 Commonly used matrices for FAB–MS.

Matrix selection and preparation for MALDI–MS was initially as simple as the FAB method. However, the introduction of new matrix materials for MALDI–MS has complicated matrix selection, at the same time having a tremendous impact on sensitivity and accuracy. For instance, it is now possible to observe peptides with attomole sensitivity (Chapter 4).

Sample preparation with MALDI involves dissolving the sample into a solvent, such as water, acetone, or methanol. The sample solution is then added to the matrix—commonly a nicotinic acid or cinnamic acid derivative—at a ratio of approximately 1 part sample to 10,000 parts matrix. Approximately 10 pmol (1 μl of a 10 μM solution) of the sample is added to the matrix. It is important to note that, unlike FAB analysis where a liquid matrix of NBA or glycerol is used, MALDI most commonly involves a solid matrix. Therefore, MALDI sample preparation requires that after the sample solution is made up and deposited upon the probe, it is necessary to wait for complete evaporation of the solvent to leave a solid solution of the sample in the matrix.

Figure 3.5 The single sample probe is common for FAB analysis. The multisample probe is more common for MALDI.

Because a sample cannot be analyzed until the solvent completely evaporates, sample preparation for MALDI will usually take longer than FAB preparation. The evaporation process can be expedited by flowing a stream of nitrogen or air over the dissolved matrix or by using a volatile solvent (acetone) if possible. Multisample probes also quicken the analysis time by allowing for many samples to be prepared at once (Figures 3.5 and 3.6). In fact, multisample probes are now used for combinatorial library work, where high sample throughput is required.

Some of the more common matrix materials for MALDI–MS are shown in Figure 3.7.

Salt Purification

Whether analysis is being performed with FAB, MALDI, or ESI, salt content should be minimized. While in some cases salts will facilitate ionization for certain compounds, in general excessive salt contamination will lead to reduced sensitivity. This is also true of chaotropic agents, including urea and guanidinium salts, and solvents like dimethyl sulfoxide and glycerol. If these agents are necessary for your work, it is then necessary to purify the compounds before mass analysis. Dialysis and reverse-phase liquid chromatography, or exchange chromatography, are preferable purification methods followed by concentration.

Figure 3.6 Difference between MALDI and FAB preparation.

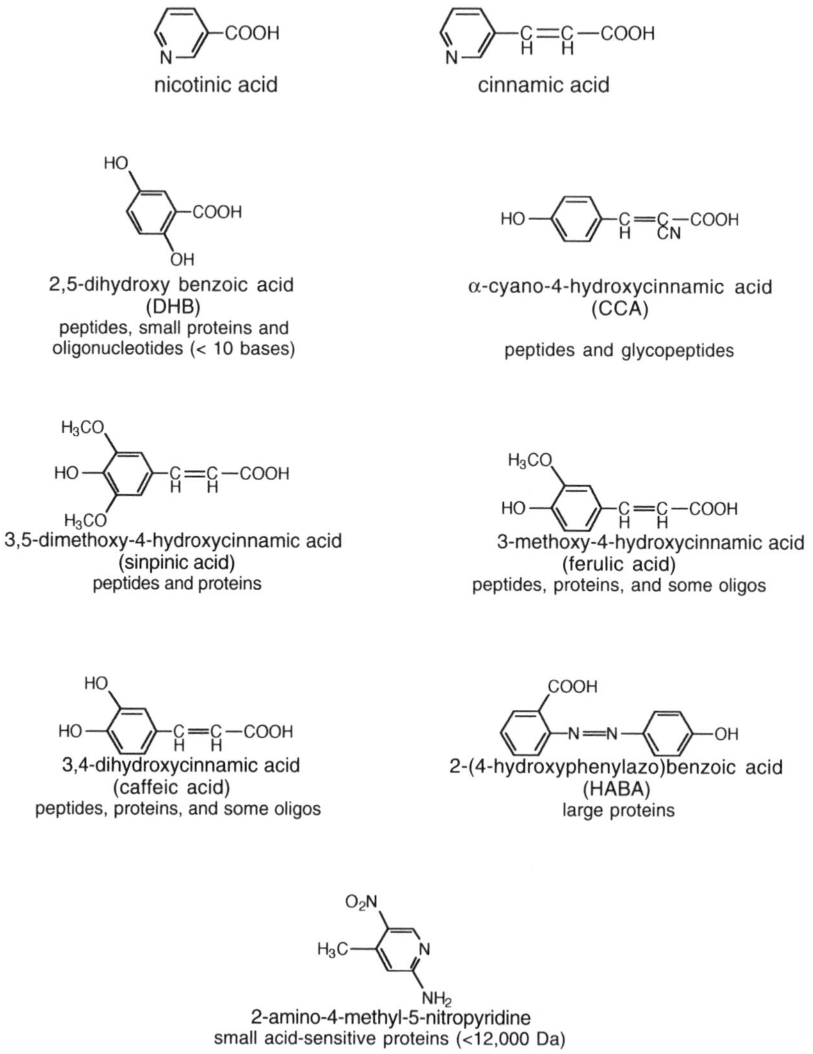

Figure 3.7 Commonly used matrices for MALDI–MS.

Calculating Molecular Weight

Besides questions concerning sample quantity and quality, another question that often arises is how to calculate molecular weight before a sample is even analyzed. There are three different ways to calculate mass from the molecular formula: (1) the average mass, (2) the monoisotopic mass, and (3) the nominal mass. Each is defined below.

Average mass–*The mass of an ion for a given empirical formula, calculated using the average atomic weight, averaged over all isotopes, for each element. For* $C_{60}H_{122}N_{20}O_{16}S_2$, *average mass = 1443.8857.*

Monoisotopic mass (exact mass)–*The mass of an ion for a given empirical formula calculated using the* exact *mass of the most abundant isotope of each element. For* $C_{60}H_{122}N_{20}O_{16}S_2$, *monoisotopic mass = 1442.8788.*

Nominal mass–*The mass of an ion with a given empirical formula calculated using the* integer *mass of the most abundant isotope of each element, for* $C_{60}H_{122}N_{20}O_{16}S_2$, *nominal mass = 1442.*

Each calculation is used under specific conditions. Average mass is used when the individual isotopes of a molecular ion cannot be distinquished (as in Figure 3.8, resolution = 200). The monoisotopic mass is used when the isotopes can be distinguished (Figure 3.8, resolution = 2000). The nominal mass is usually applied to compounds containing the elements C, H, N, O, and S, below a mass of 1000 Da when the isotopes can be distinguished. Examples are shown in Table 3.3.

The single most important point to consider when calculating mass is the resolving power of the instrument. If the resolving power is high enough and the isotopes can be distinguished, the monoisotopic mass is a satisfactory description. For instance, magnetic sector instruments will often allow

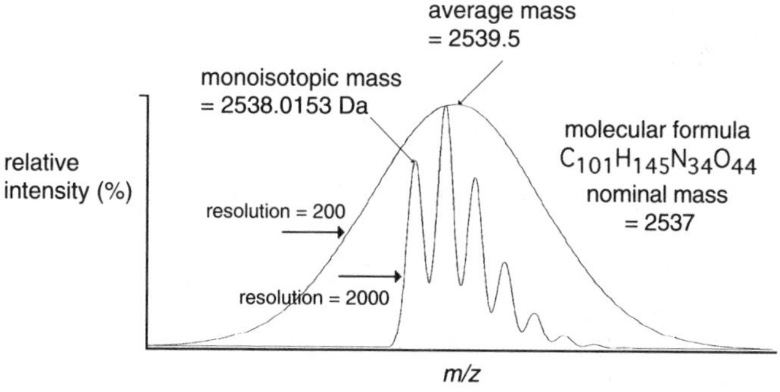

Figure 3.8 The mass spectrum of a peptide generated at resolutions of 200 and 2000 (5% valley definition). An example of how the resolution affects the observed *m/z*. At resolution 2000 all of the isotopes are resolved.

TABLE 3.3
Illustration of the Differences between Average, Monoisotopic, and Nominal Mass for Some Elements, a Lipid, a Sugar, and a Peptide

Name	Molecular formula	Average mass (Da)	Monoisotopic mass (Da)	Nominal mass (Da)
Carbon	C	12.1115	12.0000	12
Hydrogen	H	1.0080	1.0078	1
Nitrogen	N	14.0067	14.0031	14
Oxygen	O	15.9994	15.9949	16
Sulfur	S	32.0600	31.9721	32
A lipid	$C_{18}H_{35}N_1O_1$	281.4858	281.2718	281
A sugar	$C_{56}H_{118}N_4O_{14}$	1071.5833	1070.8644	1070
A peptide	$C_{101}H_{258}N_{24}O_{24}$	2193.3288	2191.9704	2190

sufficient resolving power for small molecule analysis to observe isotopes. If, however, the instrument is incapable of resolving the isotopes, the average mass will be observed and therefore should be calculated. MALDI with a time-of-flight analyzer, for example, will typically provide low resolution on the order of 100–400, so it is safest to calculate the average mass when using this technique. In general, calculating the average mass is satisfactory for compounds above a mass of 5000. For compounds below a mass of 5000, the resolution becomes an important factor in determining whether to calculate the average or monoisotopic mass.

Figure 3.8 illustrates how the isotopes of a particular compound and the resolution of the instrument can affect the mass spectrum. An instrument with a resolution of 2000 is capable of resolving all the isotopes. However, at a resolution of 200, no isotopes are resolved and the mass is observed at *m*/*z* 2539.5 (see Figure 3.8). Figure 3.8 also demonstrates the affect isotopes can have on the observed molecular weight. Compounds containing ^{12}C, ^{1}H, ^{14}N, and ^{16}O have isotopes (^{13}C, ^{2}H, ^{15}N, ^{17}O, and ^{18}O) in relatively low abundance (1.10, 0.015, 0.366, 0.038, and 0.200%, respectively), as shown in Table 3.4. While these isotopes do not make a significant contribution until masses of greater than 1000 Da are reached, other elements have significant isotopic contributions and will significantly alter the isotopic distribution (Example 3.1).

Chlorine and bromine are the two most common of these constituent elements. Looking at their isotopic patterns in Figure 3.9, note that a

TABLE 3.4
A Partial Listing of the Mass of Some Elements and Their Isotopes, Mass, and Relative Abundance[a]

Element	Mass	%	Element	Mass	%	Element	Mass	%
$^{1}\mathbf{H}$	1.0078	99.985	$^{39}\mathbf{K}$	38.9637	93.2581	$^{79}\mathbf{Br}$	78.9183	50.69
^{2}H	2.0141	0.015	^{40}K	39.9640	0.0117	^{81}Br	80.9163	49.31
$^{6}\mathbf{Li}$	6.0151	7.5	^{41}K	40.9618	6.7302	$^{127}\mathbf{J}$	126.9045	100.00
^{7}Li	7.0160	92.5	$^{40}\mathbf{Ca}$	39.9626	96.941	$^{133}\mathbf{Cs}$	132.9054	100.00
$^{12}\mathbf{C}$	12.0000	98.90	^{42}Ca	41.9586	0.647	$^{130}\mathbf{Ba}$	129.9063	0.106
^{13}C	13.0034	1.10	^{43}Ca	42.9588	0.135	^{132}Ba	131.9050	0.101
$^{14}\mathbf{N}$	14.0031	99.634	^{44}Ca	43.9555	2.086	^{134}Ba	133.9045	2.417
^{15}N	15.0001	0.366	^{46}Ca	45.9537	0.004	^{135}Ba	134.9057	6.592
$^{16}\mathbf{O}$	15.9949	99.762	^{48}Ca	47.9525	0.187	^{136}Ba	135.9046	7.854
^{17}O	16.9991	0.038	$^{55}\mathbf{Mn}$	54.9380	100.00	^{137}Ba	136.9058	11.23
^{18}O	17.9992	0.200	$^{54}\mathbf{Fe}$	53.9396	5.80	^{138}Ba	137.9052	71.70
$^{19}\mathbf{F}$	18.9984	100.00	^{56}Fe	55.9349	91.72	$^{185}\mathbf{Re}$	184.9530	37.40
$^{23}\mathbf{Na}$	22.9898	100.00	^{57}Fe	56.9354	2.20	^{187}Re	186.9558	62.60
$^{24}\mathbf{Mg}$	23.9850	78.99	^{58}Fe	57.9333	0.28	$^{184}\mathbf{Os}$	183.9525	0.02
^{25}Mg	24.9858	10.00	$^{59}\mathbf{Co}$	58.9332	100.00	^{186}Os	185.9539	1.58
^{26}Mg	25.9826	11.01	$^{58}\mathbf{Ni}$	57.9353	68.27	^{187}Os	186.9558	1.60
$^{27}\mathbf{Al}$	26.9815	100.00	^{60}Ni	59.9308	26.10	^{188}Os	187.9559	13.30
$^{28}\mathbf{Si}$	27.9769	92.23	^{61}Ni	60.9311	1.13	^{189}Os	188.9582	16.10
^{29}Si	28.9765	4.67	^{62}Ni	61.9283	3.59	^{190}Os	189.9585	26.40
^{30}Si	29.9738	3.10	^{64}Ni	63.9280	0.91	^{192}Os	191.9615	41.00
$^{31}\mathbf{P}$	30.9738	100.00	$^{63}\mathbf{Cu}$	62.9296	69.17	$^{196}\mathbf{Hg}$	195.9658	0.15
$^{32}\mathbf{S}$	31.9721	95.02	^{65}Cu	64.9278	30.83	^{198}Hg	197.9668	10.10
^{33}S	32.9715	0.75	$^{64}\mathbf{Zn}$	63.9291	48.60	^{199}Hg	198.9683	17.00
^{34}S	33.9679	4.21	^{66}Zn	65.9260	27.90	^{200}Hg	199.9683	23.10
^{36}S	35.9671	0.02	^{67}Zn	66.9271	4.10	^{201}Hg	200.9703	13.20
$^{35}\mathbf{Cl}$	34.9689	75.77	^{68}Zn	67.9248	18.80	^{202}Hg	201.9706	29.65
^{37}Cl	36.9659	24.23	^{70}Zn	69.9253	0.60	^{204}Hg	203.9735	6.80

[a] A complete listing can be found in the Appendix.

compound containing one chlorine atom will have an isotopic contribution that correlates to a peak having a mass at **M** + 2, with a height of ~32% of the primary ion. The percentage contribution for chlorine and bromine will change if more of these halogens are added to the compound (Figure 3.9). Observing these isotopes can be a useful confirmation of their presence in a particular compound.

Example 3.1: The isotopic pattern of an arginine-based compound containing two chlorine atoms.

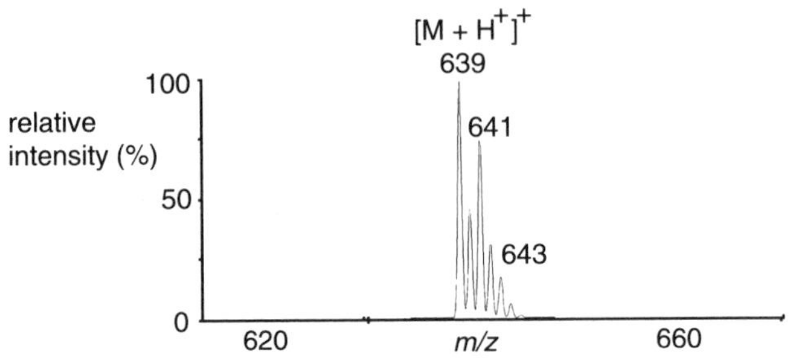

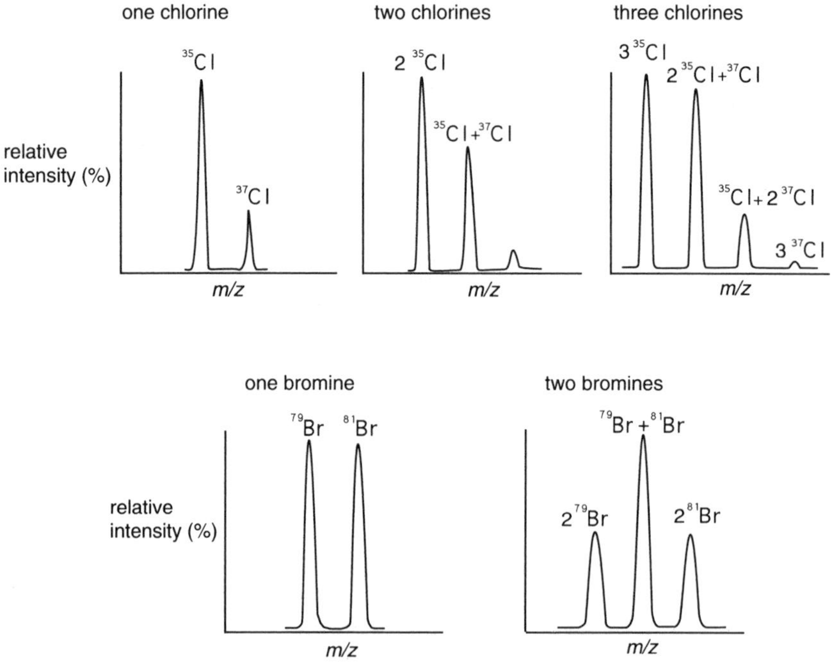

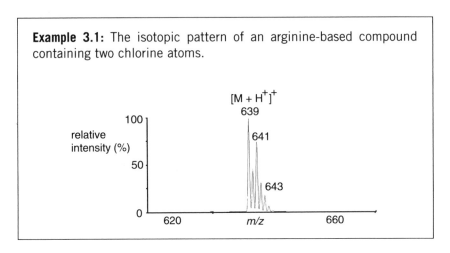

Figure 3.9 The isotopes associated with chlorine and bromine produce characteristic isotopic patterns that readily help one to identify these elements in a compound.

Quantifying

Users of mass spectrometry commonly ask, "Do the ion intensities observed in the mass spectrum correlate to the relative amounts of each component?" They are usually given a negative answer, unless the compound has been carefully prepared and calibrated against an internal standard. However, for compounds that are similar in mass and have the same functional groups, the relative ion intensities may qualitatively correspond to their content. Ionization efficiency is a critical issue with regard to quantifying relative sample concentration. If two compounds have very similar ionization efficiencies and are of similar mass, their ion intensities may qualitatively correlate to the relative quantity of each sample. However, this is not true for two compounds of similar mass having significantly different ionization efficiencies. For example, in positive FAB mass spectrometry with equal amounts of two compounds with a single functional group, an amine and an amide, the amine would dominate the mass spectrum because of its relatively high proton affinity and correspondingly high ionization efficiency (Examples 3.2 and 3.3).

Liquid Chromatography–Mass Spectrometry

In the past, attempts were made to couple liquid chromatography with mass spectrometry. These attempts did produce some success. However, electrospray ionization has made LC–MS routine. The ability of electro-

Example 3.2: An amine and an amide have different proton affinities and it is not possible to quantitatively compare their content from the ion intensities. The spectrum shown is from an equal amount of each compound, illustrating that the ion intensity does not necessarily correlate to the amount of sample being analyzed.

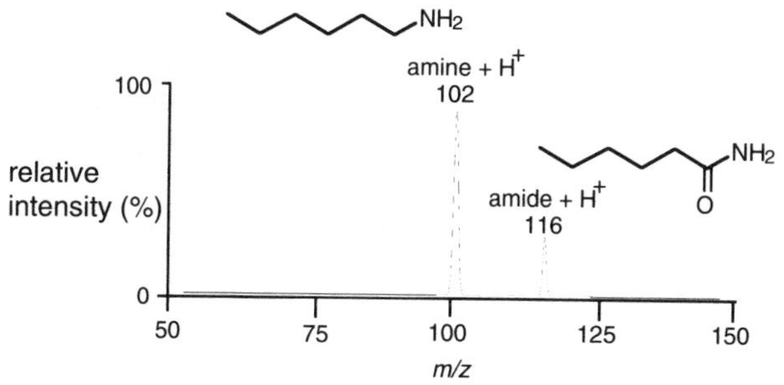

Example 3.3: Our laboratory has pursued the quantitative anlaysis of cyclosporin A (an immunosuppressant), using cyclosporin G as an internal standard. These two molecules differ by a single methyl group. The similarity in structure between the two molecules makes their ionization properties, for practical purposes, identical. Therefore, adding cyclosporin G (at 0.20 μM) as an internal standard into the solution containing cyclosporin A allowed us to produce the linear relationship between the two using electrospray mass analysis.

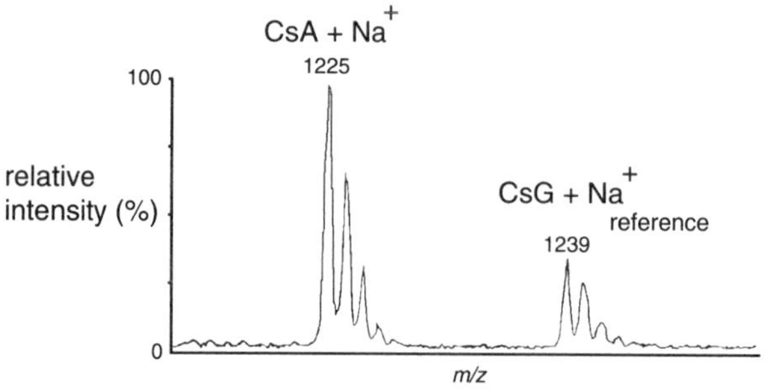

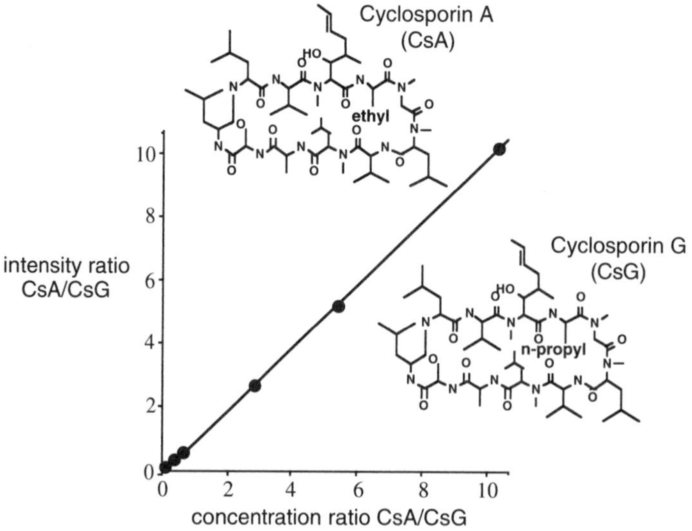

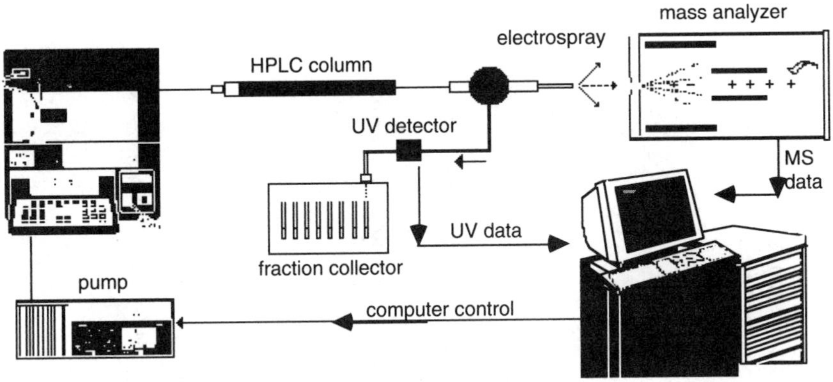

Figure 3.10 Interfacing liquid chromatography with ESI mass spectrometry.

spray ionization mass spectrometry to directly analyze compounds from aqueous or aqueous/organic solutions has established the technique as a convenient mass detector for liquid chromatography. Figure 3.10 illustrates how mass spectrometry can serve as a detector to liquid chromatography.

Capillary Zone Electrophoresis–Mass Spectrometry (CE–MS)

Electrospray ionization has also been interfaced with capillary electrophoresis. CE–MS is particularly well suited for the analysis of large molecules because of the multiple charging phenomenon and the low flow rates associated with CE.

Table 3.5 presents some common questions about mass spectrometry and the different ionization techniques with the corresponding answers.

TABLE 3.5
Common Questions Asked about Mass Spectrometry and Biomolecule Applications

Question	Answer
1. What concentration do you need?	20–50 μM for electrospray 20–50 μM for MALDI 1000 μM for FAB
2. How much solution do you need (given the concentrations above)?	≤50 μl for electrospray ≤5 μl for MALDI ≤5 μl for FAB
3. How high in molecular weight can you go?	Typically ~300,000 Da by MALDI ~70,000 Da by electrospray ~7,000 Da by FAB
4. What does resolution mean in mass spectrometry?	Resolution is the ability of an instrument to distinguish ions of differing *m/z*. High resolution enables the spectroscopist to resolve isotopic peaks of high-mass ions. The simplest definition uses the formula

$$\text{Resolution} = M/\Delta M = 500/0.5 = 1000$$

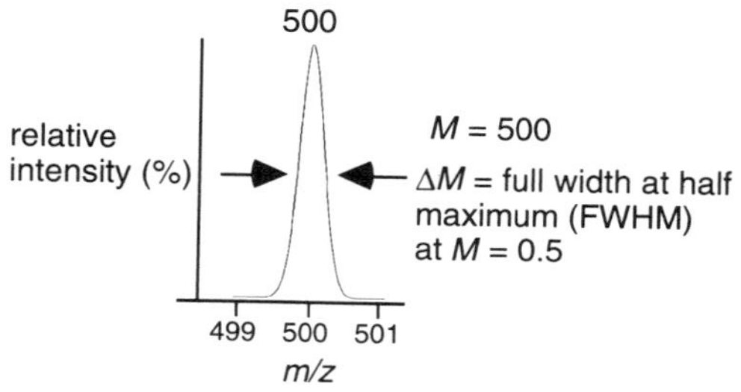

5. Do you have to have matrix peaks in the FAB or MALDI mass spectra?	Yes. Depending on the instrument, it is possible to subtract out the matrix signal when processing the data. In general, these peaks can be an informative source of accuracy and sensitivity for the spectroscopist and are usually kept in the spectrum.
6. Are there matrix peaks in the electrospray mass spectra?	No. Since electrospray uses solvents like water, acetonitrile, methanol, or chloroform, no typical matrix peaks are observed. However, it is common to observe aggregates of water or other solvents, and also common to see salt clusters, especially at high salt concentrations. Typical solvent clustering will be significant only for $m/z \leq 300$.
7. Why are Na^+ or Cs^+ sometimes added to the matrix or solvent solution?	Many compounds do not form stable molecular ions when they are protonated. This is because the charge from protonation can be transferred onto the molecule, which may destabilize the molecular ion. Alkali cations are therefore added to the sample solution, because once the cation is complexed to the molecule, the charge remains localized on the cation and does not transfer onto the molecule. Thus, there is no charge destabilization. Also, the salts combined with the matrix will often give useful reference ions. These reference ions can make it easier to perform exact mass measurements or to check the calibration.
8. Why don't you *always* add NaI or CsI to the matrix or electrospray solution?	Compounds with acidic functional groups are especially sensitive to the presence of salts. Because salts can make it difficult to observe the molecular ion, we try not to use them with acidic compounds, or use as little salt as possible. In addition, the presence of salts can affect the ionization efficiency in FAB, MALDI, and electrospray. In general, it is good practice to minimize salt contamination. Salts can always be added, but removing them is more difficult.
9. Is there any way to tell if a peak in electrospray is singly or doubly charged?	Look at the isotope pattern. If the spacings between the isotopes are 1 mass unit apart, the ion is singly charged. $\frac{1}{2}$ mass unit spacing means double charge, $\frac{1}{3}$ mass unit spacing means triple charge. This is illustrated in Chapter 4. Most quadrupole instruments will not resolve more than the triply charged isotopic patterns.

(continued)

TABLE 3.5 (*Continued*)

Question	Answer
10. Why is electron ionization not used as often as it used to be?	Electron ionization has many limitations (Chapter 1), and because of these limitations many compounds (especially thermally labile compounds) will not work using electron ionization. FAB, MALDI, and electrospray work much more consistently than electron ionization on biological compounds.
11. Why. does salt cause problems in electrospray ionization?	Electrospray ionization forms ions by the evaporation of charged droplets. One possible reason why salt can decrease the formation of ions is that its presence inhibits the evaporation process, and therefore the ionization process as well. All the effects of salt on ionization are not understood. However, it is known that excessive salt ($>500\ \mu M$) is often counterproductive to obtaining high-quality electrospray mass data.
12. Why are mixtures or heterogeneous compounds so difficult to analyze?	FAB, MALDI, and electrospray ionization generate ions, usually by the competitive addition of a cation. If there is more than one component in the solution, the compound with the highest affinity for the cation will usually dominate the spectrum. Impurities such as salts and detergents will also affect the ionization process in the same way or by minimizing the ion's ability to get to the surface of the solution. These impurities will also affect the microcrystallization of the MALDI matrix necessary for efficient ionization.
13. Are very volatile samples easy or hard to analyze?	Volatile samples are neither easy nor hard to analyze. Their volatility is a factor that usually determines which ionization technique to use. For instance, a very volatile compound may not work well with FAB, MALDI, or electrospray and may require electron ionization as an ionization source.
14. Why do you need to know the structure of a compound to do mass analysis?	The structure or general idea of what the compound is can be very useful. For routine mass analyses the structure is very informative. It gives us an idea of what ionization technique to use, what matrix/solvent to use, whether to analyze in a positive or negative ionization mode, how much sample we may need, and it affords a learning experience for the person analyzing the sample. The spectroscopist can learn from every sample he or she analyzes. However, nothing is gained if the structure is not presented.

CHAPTER 4

Peptide and Protein Analysis

This chapter describes peptide and protein analysis with the FAB, MALDI, and electrospray ionization techniques as well as peptide and protein sequencing techniques.

Perspective

As we push ahead in our investigation of mass spectrometry, it seems appropriate to explore peptides and proteins, which have been driving forces in the field. The word "protein" derives from the Greek *proteios,* meaning "of first importance." The scientific importance of characterizing these compounds was not lost on mass spectoscopists. The FAB, MALDI, and electrospray ionization techniques have had a significant impact on peptide and protein mass analysis, with four major goals realized:

Routine Molecular Weight Determination of Peptides–FAB, MALDI, and electrospray have been successful in routine peptide analysis.

Routine Molecular Weight Determination of Proteins–MALDI and electrospray have been successful in routine protein analysis (ESI to 70 kDa and MALDI to 300 kDa).

High Sensitivity–The ability to observe subfemtomole (attomole) amounts of peptide is now possible with MALDI, while electrospray is useful down to femtomole quantities of peptides. MALDI and electrospray allow for picomole protein analysis.

Structural Information on Peptides–Structural information on very small quantities of peptide has been obtained. The current benchmark, set by Don Hunt, is the characterization of major histocompatibility complex (MHC) peptides at the femtomole level using electrospray ionization.

The FAB, MALDI, and ESI methods are all effective in generating peptide ions, and the results are straightforward. Peptides form cations [peptide + H]$^+$, [peptide + 2H]$^{2+}$, or [peptide + nH]$^{n+}$, as shown in Figure 4.1, and can also be accompanied by some fragmentation for peptides of less than 20 amino acids. In addition, mass spectrometry has also demonstrated its utility in identifying chemically modified peptides, as these modifications usually lead to a predictable change in molecular weight.

Protein ionization is achieved by the addition of a proton or protons. Because of a protein's larger size, however, the sensitivity for its analysis is typically lower. Even so, the sensitivity of MALDI and electrospray for proteins is still considered good at the picomole level or even hundreds of femtomoles. Partial structural identification of proteins is now achieved by chemical degradation followed by mass analysis of the decomposition products.

Ultimately, scientists hope to use mass spectrometry to fully sequence both peptides and proteins. While this is done to some extent now on peptides, routine, rapid, and complete sequence determination through mass analysis of proteins is yet to be realized. This chapter focuses on the current analysis capabilities in peptide and protein mass spectrometry using FAB, MALDI, and electrospray ionization sources. The specific preparation procedures, matrix selections, and applications of each method will be discussed, along with the application of tandem mass spectrometry toward sequencing peptides and proteins.

Peptide and Protein Analysis by FAB

The application of FAB to peptide analysis has two distinct advantages: it typically generates accurate molecular weight information and offers rapid analysis. Since FAB has a limited mass range, routinely on the order of 5000–8000 Da, and relatively poor sensitivity (typically requiring >100 pmol), its application to protein analysis is not common. With that

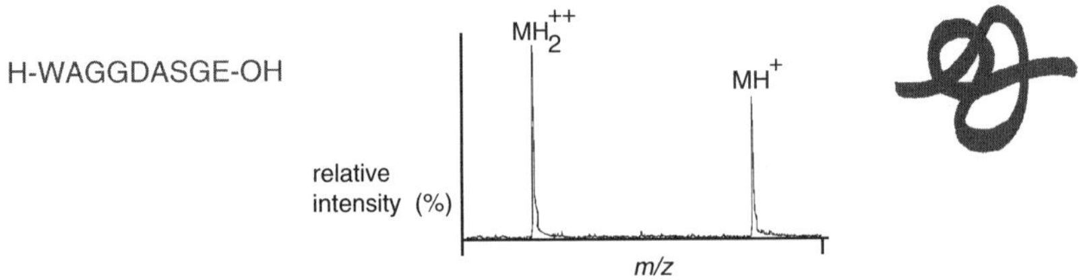

Figure 4.1 Example of peptide mass spectral data obtained from an electrospray mass spectrometer.

in mind, the following discussion focuses on the application of FAB to peptide analysis.

FAB peptide preparation is accomplished by dissolving a relatively pure peptide (>70%) sample into a *m*-nitrobenzyl alcohol matrix. The sample can be prepared in a cosolvent at a concentration of approximately 1–5 mg/ml. Sample solubility in this cosolvent is crucial because it acts as a medium between the sample and the matrix. Two to 4 μl of the peptide solution is dissolved directly into the matrix on the probe tip (a flat round surface approximately 2–5 mm in diameter), as shown in Figure 4.2. When large amounts of sample (hundreds of micrograms) are available, the peptide can be added directly into the matrix, followed by the addition of a cosolvent like triflouroacetic acid (TFA) which, in addition to sample matrix solubility, can also facilitate protonation.

A sample preparation procedure for FAB is described below:

Matrix: place 2–5 μl of NBA directly on the probe.
Sample: dissolve sample in 50:50 acetonitrile:water with 0.1% trifluoroacetic acid. Sample concentration should be ~5 μg/μl.
Add 1–4 μl of sample solution to the matrix.
Wait approximately 1 min for solvent to evaporate.
Acquire spectra.

There are many matrices used in FAB analyses; however, *m*-nitrobenzyl alcohol has become a standard for FAB peptide analysis (Figure 4.3). Cosolvents include 5–50% acetic acid in water, acetonitrile/water, chloroform, methanol, and triflouroacetic acid (0.1–25%).

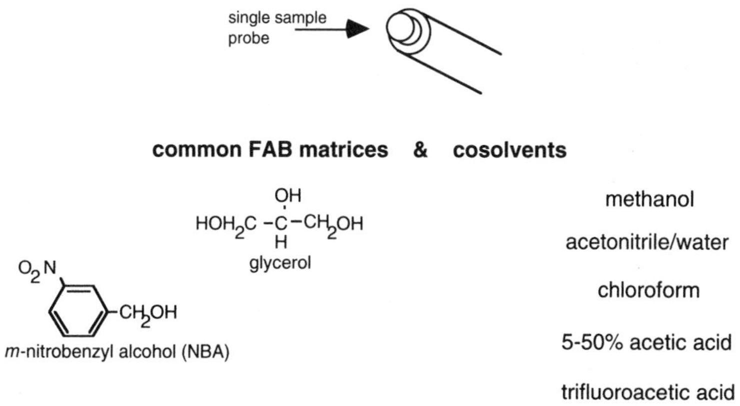

Figure 4.2 An example of the FAB single-sample probe and some of the matrices and cosolvents used in FAB mass spectrometry of peptides.

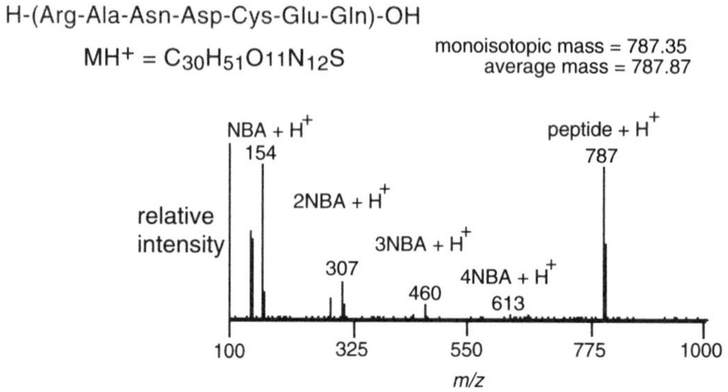

Figure 4.3 The FAB mass spectrum of a peptide analyzed in the NBA matrix. Notice the peaks associated with this matrix. These matrix peaks can make the spectra more cumbersome to interpret. However, to the experienced mass spectrometrist the matrix ions can act as reference ions and are a useful indicator of accuracy (rarely are the NBA matrix ions observed above *m/z* 1000).

Recently, FAB has demonstrated some success with the mass analysis of large peptides using massive cluster impact (MCI). Massive cluster impact (Figure 4.4) uses glycerol ion clusters for particle bombardment instead of the more traditional Cs^+ ions or Xe atoms, as Cs^+ and Xe desorption is typically limited to proteins of <7000 Da. Massive cluster ionization has demonstrated an increased mass range through multiple charging of a 17,000-Da peptide and, as it is interchangeable with current Cs^+ and Xe

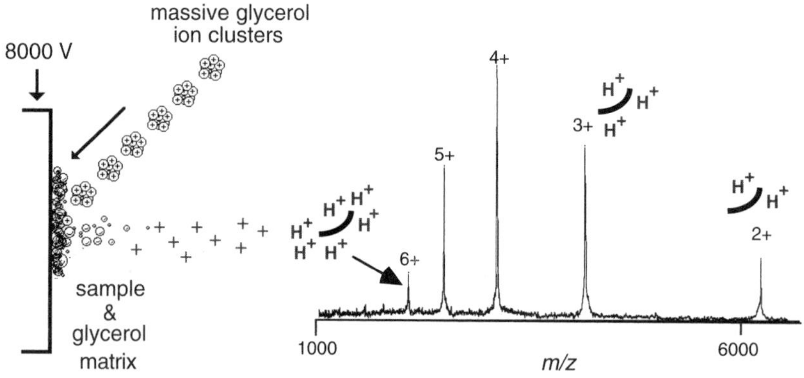

Figure 4.4 FAB mass spectrum of cytochrome *c* obtained on a JOEL HX100HF double-focusing magnetic sector instrument using massive cluster ionization (MCI). (Adapted from data presented by Phrasor Scientific Inc.)

FAB sources, MCI has demonstrated its utility for quadrupole and magnetic sector instruments which already have FAB analysis capabilities. While MCI is still unproven for routine biopolymer analysis, it may become a practical (although less sensitive) alternative to electrospray.

Since the original application of FAB, new instrument designs and matrices have made FAB a more powerful tool for the routine analysis of peptides. In the following sections, however, we will see that MALDI and electrospray have a distinct advantage in sensitivity and mass range.

Peptide and Protein Analysis by MALDI

Laser desorption of peptides and proteins became possible in the 1980s through the work of Franz Hillenkamp and Michael Karas (M. Karas, D. Bachmann, U. Bahr, and F. Hillenkamp, *Int. J. Mass Spectrom. Ion Proc.*, 1987, **78,** 53–68). It was accomplished by preparing samples in a large excess of a radiation-absorbing matrix, hence the name matrix-assisted laser desorption ionization (MALDI). MALDI mass analysis of peptides and proteins is typically accomplished with a time-of-flight analyzer which, with resolving capabilities on the order of 200–500, as demonstrated in Figure 4.5, represents MALDI's biggest disadvantage (Chapter 2). Typical accuracies for MALDI range from ±0.5 to 0.01% depending upon the presence of an internal standard, the type of instrument being used, and the selection of matrix material.

MALDI is useful for peptide analysis in three ways: (1) it allows for rapid preparation and analysis, (2) it is very sensitive, and (3) it is tolerant of

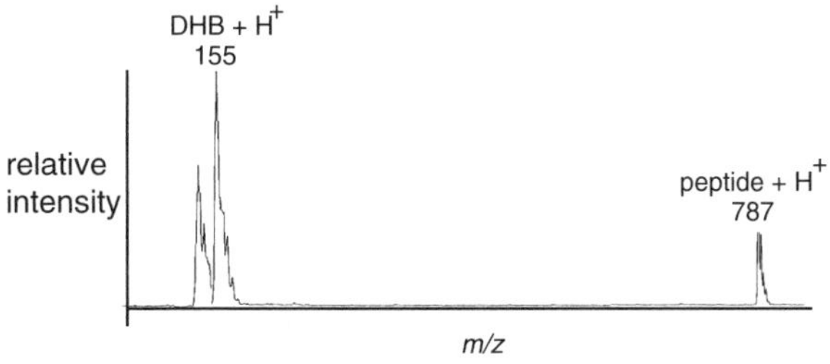

Figure 4.5 The MALDI mass spectrum using 2,5-dihydroxybenzoic acid (DHB) matrix material and the same peptide analyzed in Figure 4.4 by FAB–MS. The primary difference in the two spectra is the width of the peaks. The MALDI peaks are much broader because of the relatively low resolution of time-of-flight mass analyzers. Low resolution becomes more apparent at higher mass and effects the accuracy. MALDI combined with time-of-flight analysis makes up for its low resolution with high sensitivity and a relatively high tolerance of heterogeneous samples.

heterogeneous samples. The use of multisample probes also makes MALDI useful for preparing many samples simultaneously (Figure 4.6). Beginning with preparation, the sample preparation procedure described below (thin layer method for peptides) offers an appropriate example of MALDI's sensitivity. The following procedure from Vorm *et al.* (1994) has yielded attomole sensitivity, as seen in Figure 4.7.

Matrix: prepare a saturated solution of α-cyano-4-hydroxycinnamic acid (CCA) in acetone followed by the addition of 2% water.
Clean target: sonicate target for 15 min in formic acid/ethanol/water, 1 : 1 : 1; rinse with acetone.
Spread about 0.5 μl of the matrix solution on a polished target. A shiny, colorful film must form consisting of homogeneously distributed matrix crystals (~5.0 mm in diameter).
Add 0.3–1.0 μl of sample solution (25 fmol; ~10 attomole–100 pmol possible) in acidic aqueous solution (e.g., >2% TFA, <10% acetic acid, <70% formic acid). *Important! The matrix dissolves at neutral to basic pH values. At low pH it is even resistant to ca. 20% acetonitrile.*
Let the sample dry at ambient air or allow sample adsorption for ca. 3 min before removing excess solvent using a stream of pressurized air.
Acquire spectra.
If cleanup is necessary: add 10 μl of 0.1% TFA, blow off liquid after 2–15 sec using a stream of pressurized air and, again, take spectra.

Although peptide preparation for MALDI is somewhat different from FAB, the end result is the same—a protonated molecular ion, $M + H^+$. The $M + Na^+$ and $M + K^+$ adducts are also often observed.

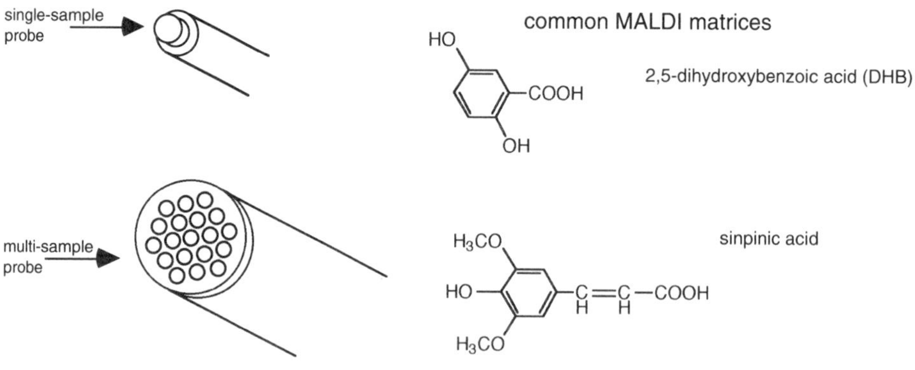

Figure 4.6 An example of the MALDI single-sample and multisample probes and two common matrices used in MALDI mass spectrometry of peptides.

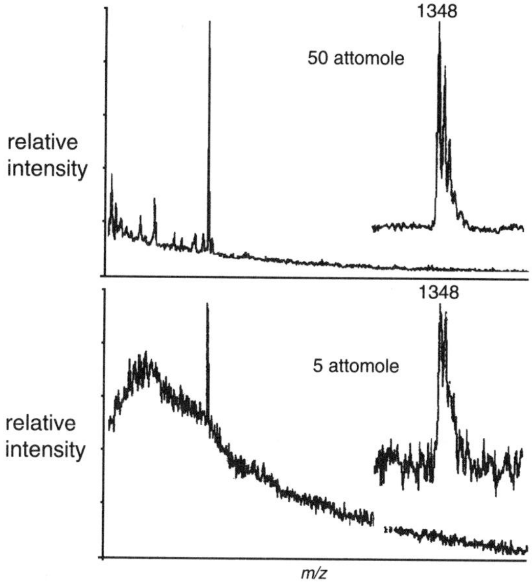

Figure 4.7 Mass spectrum of 50 and 5 attomol of a peptide (substance P seq: RPKPQQFFGLM-NH_2, monoisotopic, M_r = 1347.7) using the fast evaporation method described and a high-resolution MALDI reflectron time-of-flight mass spectrometer. This is an excellent demonstration of the accuracy and sensitivity offered by the reflectron instrumentation. Adapted from Vorm *et al.* (1994).

Concerning matrix selection, unlike FAB analysis where a liquid matrix of NBA or glycerol is used, MALDI most commonly involves a solid matrix. For MALDI analysis of small molecules and peptides (200–1000 Da), the 2,5-dihydroxybenzoic acid (DHB) matrix works well because it produces only a minimal amount of interference in the low-molecular-weight range (Figure 4.5); other matrices can give a high background of matrix signals in this mass range. Most matrices reported to date have been acidic, but basic matrices have also been introduced, such as the 2-amino-4-methyl-5-nitropyridine matrix, which extends the utility of MALDI to acid-sensitive peptides, proteins, or other acid-sensitive compounds (Fitzgerald *et al.*, 1993).

In addition to sensitivitity, MALDI has other useful features and capabilities. These include an ability to measure peptides and other compounds in the presence of salt and the ability to analyze complex mixtures. A dramatic illustration of mixture analysis is shown below with protein ladder sequencing, where a mixture of peptides/proteins is formed by Edman degradation. MALDI mass analysis of the degradation mixture then allows

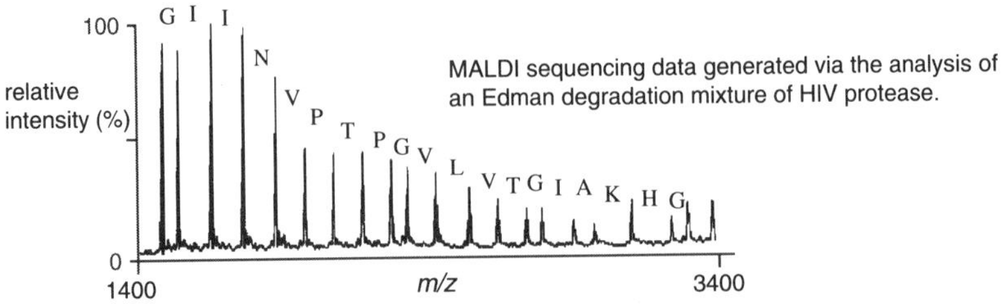

Figure 4.8 Mass spectrum of a mixture of peptides following sequential edman degradation of a protein. Adapted from Chait *et al.* (1993).

for sequence determination, as shown in Figure 4.8. MALDI can also be applied to the analysis of peptides generated from digests of a protein—a procedure used for peptide mapping.

Many of the advantages that MALDI offers for peptide analysis are equally applicable to proteins. Protein analysis is similar to peptide analysis, in which ionization usually occurs through the addition of one, two, or three protons. However, as stated in the beginning of the chapter, since proteins are significantly bigger than peptides, ion detection is less efficient. Therefore, while peptides are measured at the femtomole or even attomole level with MALDI, proteins are usually measured at the high femtomole to low picomole level. A typical spectrum of bovine serum albumin is shown in Figure 4.9.

A preparation protocol for proteins, developed by Chait and coworkers (Martin de Llano *et al.,* 1993), is described as follows.

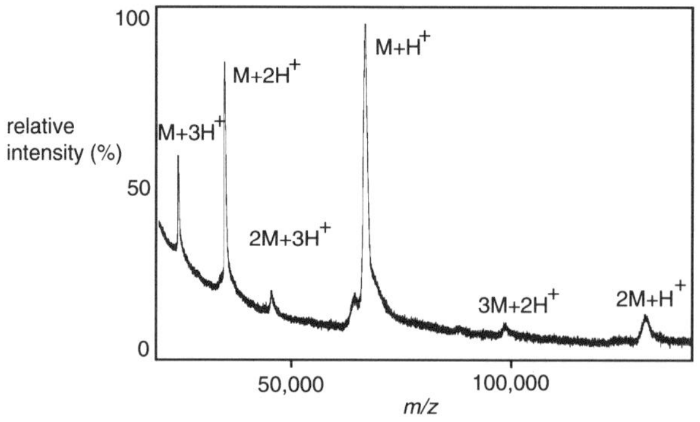

Figure 4.9 MALDI–MS of bovine serum albumin protein (BSA). Spectrum courtesy of Kenway Hoey, R. W. Johnson Pharmaceutical Research Institute.

Matrix: prepare saturated solution of CCA in 2-propanol/water/formic acid, 2:3:1 solution.

Peptides: mix 1 μl peptide solution (1–10 pmol/μl) in 0.1% TFA with 9 μl matrix solution. Apply 2 μl onto the sample target.

Proteins: mix 5 μl protein solution (1–10 pmol/μl) in 0.1% TFA with 5 μl matrix solution. Apply 1 μl onto the sample target.

Use a gentle stream of cold air from a blower to assist sample drying.

Washing the sample once it has formed a solid solution with the matrix on the probe serves to remove contaminants like salts and detergents. An example of washing the sample after it has been deposited on the matrix is shown in Figure 4.10.

MALDI is especially useful for chemically modified proteins. For instance, MALDI is currently the most reliable technique for analyzing glyco-

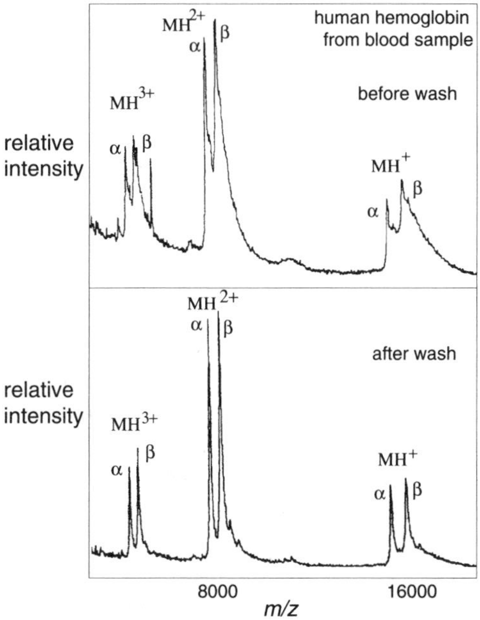

Figure 4.10 MALDI mass spectra of human hemoglobin α- and β-chains obtained before (top) and after washing (bottom). The hemoglobin was obtained from a fresh blood sample of 0.3 μl that was diluted to 200 μl in water and sonicated for ~7 min. This solution was then further diluted by a factor of 15 in 10% formic acid, and 0.5 μl of that solution was added to a matrix surface of α-cyano-4-hydroxycinnamic acid. (Adapted from Vorm *et al.,* 1994).

proteins, where the extreme broadness in their peaks reflects the carbohydrate heterogeneity in the proteins.

The determination of a glycoprotein primary structure involves determining the amino acid sequence, the structure of the carbohydrate, and the carbohydrate's point of attachment. MALDI has been demonstrated as a useful tool in each of these identification steps. Amino acid sequence information can be obtained by MALDI when used in conjunction with digestion techniques, and recent reports suggest that carbohydrate structures can also be determined.

DHB is the most common matrix for glycoprotein analysis. This matrix performs well for 150 fmol amounts of smaller glycopeptides in both positive and negative ion spectra (Huberty *et al.*, 1993) and is even being used to measure antibodies.

The following procedure using 2,5-dihydroxybenzoic acid has also produced good data with MALDI for a wide variety of compounds:

Matrix: make 10 mg/ml solution of DHB in 0.1% trifluoroacetic acid in water/acetonitrile, 2:1.
Make a 10 pmol/μl solution of glycoprotein in 0.1% trifluoroacetic acid in water/acetonitrile, 2:1.
Make a 3:1–10:1 mixture of matrix and protein solution, depending on protein molecular weight.
Apply 0.5–2 μl onto sample target.
Use a gentle stream of cold air from a blower to assist sample drying.

Figure 4.11 demonstrates the ability of MALDI to handle large proteins.

Peptide and Protein Analysis by Electrospray

As with MALDI and FAB, the electrospray ionization of these compounds involves the addition of a proton or multiple protons. Sample preparation is achieved by dissolving the sample in a protic volatile solvent system that is relatively homogeneous and less than 500 μM of salt. Because of its high mass accuracy and because it can be easily interfaced with liquid chromatography, electrospray is the method of choice for peptides and low-molecular-weight proteins below 70,000 Da. A sample spectrum is shown in Figure 4.12. There are only a few preparation procedures and no matrices for electrospray. This makes electrospray a quick method for obtaining molecular weight.

In addition to high accuracy, another important feature of most electrospray mass spectrometers is their ability to generate and observe multiply charged species. Multiple charging makes it possible to observe large proteins with mass analyzers that have a relatively small mass range. In addition,

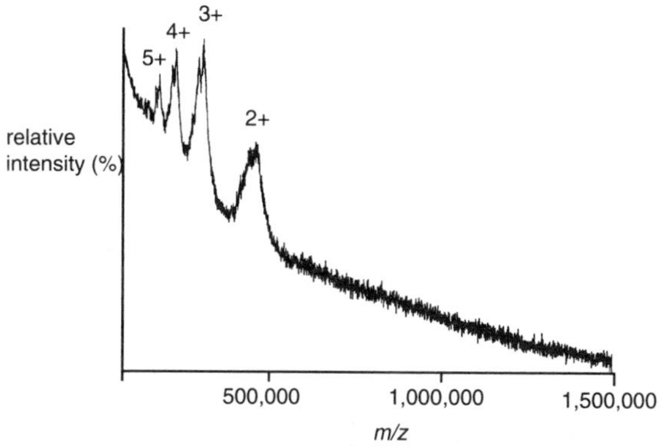

Figure 4.11 MALDI mass spectrum of a human immunoglobulin IgM protein, $M_r = 982{,}000$. The spectrum was acquired from 70 fmol of analyte using a matrix of sinpinic acid. (Courtesy of Nelson *et al.*, 1994).

observing multiple peaks for the same peptide allows one to make multiple molecular weight calculations from a single spectrum. Thus one can average these values and obtain a very accurate molecular weight.

The question often arises, especially concerning the analysis of peptides and proteins, "How is the charge state of the observed ion determined?" The answer is straightforward. In the spectrum in Figure 4.13 two peaks are seen, one at *m/z* 1978 and another at *m/z* 990. How does the investigator know there are not two separate peptides? Are you looking at one or two peptides? Since most electrospray spectrometers have good resolution, it is often possible to look at the isotopic distribution of singly and doubly charged ions.

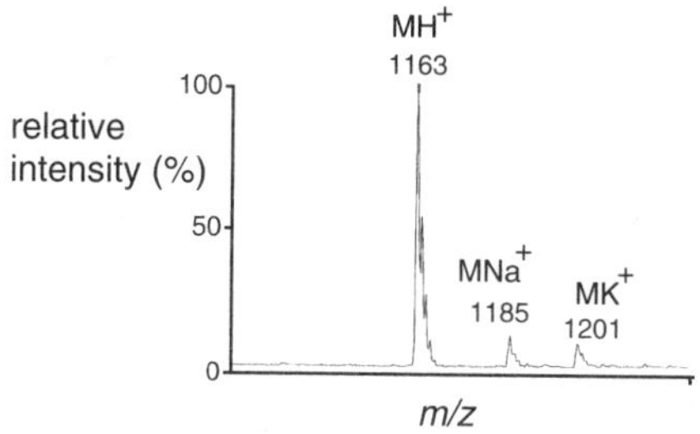

Figure 4.12 An electrospray mass spectrum of a peptide.

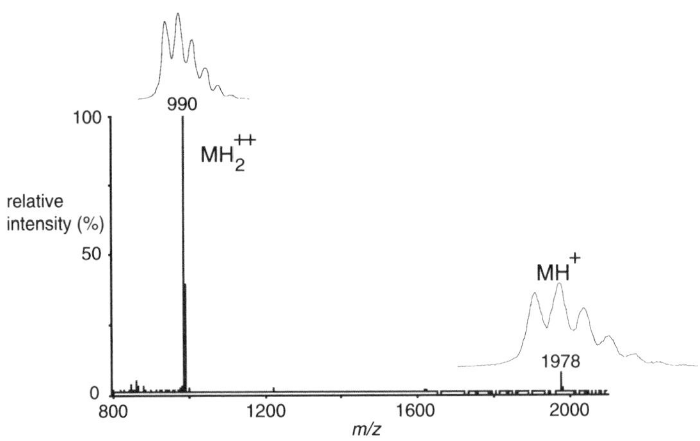

Figure 4.13 The question of whether a peptide (or any other species) observed in the mass spectrum is singly, doubly, or multiply charged can be addressed in several ways. In this spectrum it is possible to look at the isotopic distribution and determine the spacing between the isotopes. If isotopes are separated by one mass unit the charge is 1+; if by $\frac{1}{2}$ mass units, the charge is 2+.

By looking at the isotopic distribution of the above ions (Figure 4.13) we can readily see the isotopes are separated by one mass unit at *m/z* 1978 and $\frac{1}{2}$ mass units at *m/z* 990 (Figure 4.14). Each pattern therefore corresponds to the 1+ and 2+ charge states, respectively. This is due to the fact that the mass spectrometer measures the mass-to-charge ratio. Therefore, as the calculation demonstrates, for the 2+ charge state the isotopes are separated by $\frac{1}{2}$, at 3+ the isotopes are separated by $\frac{1}{3}$, and at 4+ the isotopes are separated by $\frac{1}{4}$. Unfortunately, the resolving power of the common quadrupole electrospray mass spectrometers only allows for distinguishing between singly and doubly charged species.

Fortunately, as bigger peptides and proteins are analyzed, a distribution of ions is obtained (Figure 4.14). Even though we cannot look at the individual charge states from the isotopic pattern, we can deduce the charge state by looking at a couple of peaks in the distribution. The calculations given in Chapter 1 for determining charge state were somewhat tedious. However, it is unnecessary to discuss these calculations, because all commercial instruments allow for these calculations to be performed very easily. Figure 4.15 also illustrates the results of two computer-generated calculations designated as hypermass reconstruct.

New reports on electrospray technology, published on a monthly basis, shed insight into this interesting tool. One particular report confirms that

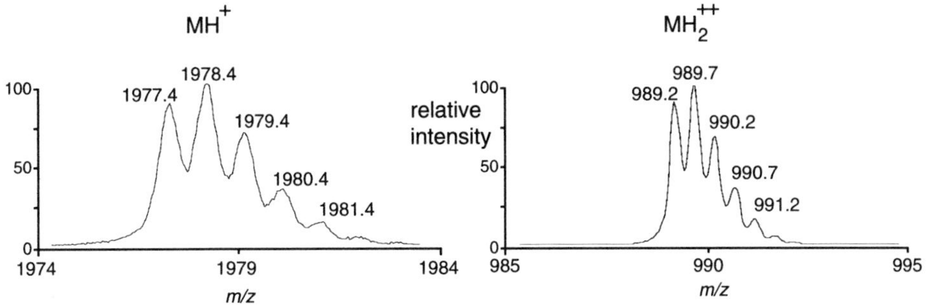

(molecular mass + #protons)/charge = mass-to-charge ratio

singly charged	doubly charged
(1976.4 + 1.0)/1 = 1977.4	(1976.4 + 2.0)/2 = 989.2
(1977.4 + 1.0)/1 = 1978.4	(1977.4 + 2.0)/2 = 989.7
(1978.4 + 1.0)/1 = 1979.4	(1978.4 + 2.0)/2 = 990.2
(1979.4 + 1.0)/1 = 1980.4	(1979.4 + 2.0)/2 = 990.7
(1980.4 + 1.0)/1 = 1981.4	(1980.4 + 2.0)/2 = 991.2

Figure 4.14 The isotopic pattern of a peptide with singly and doubly charged molecular ions obtained from an electrospray quadrupole mass spectrometer. Peptide courtesy of Anita Everson, R. W. Johnson Pharmaceutical Research Institute.

oxidation can occur during the electrospray ionization process and discusses the additional minor peak that is often observed 16 mass units above the expected mass of the peptide $[M + H^+ + 16]^+$. Apparently this peak is acquired from the oxidation of methionyl, tryptophanyl, or tyrosyl amino acids present in the peptide. The study further showed that oxidation was dependent on the voltage applied to the electrospray nozzle.

The literature shows that the analysis of hydrophobic biopolymers, like membrane peptides and membrane proteins, is infrequent. This deficiency is due to the incompatibility between mass analysis techniques and the detergents and/or salts that are required to retain such proteins in solution. Hydrophobic peptides and proteins are generally not soluble in the methanol/water or acetonitrile/water solutions typically used for electrospray, and most detergents and chaotropes that are useful for dissolving these compounds interfere with their mass analysis. Schindler *et al.* (1993) identify the use of chloroform/methanol/water mixtures as well as hexafluoroisopropanol and 70–95% formic acid as useful solvent systems with hydrophobic biopolymers, particularly bacterioopsin.

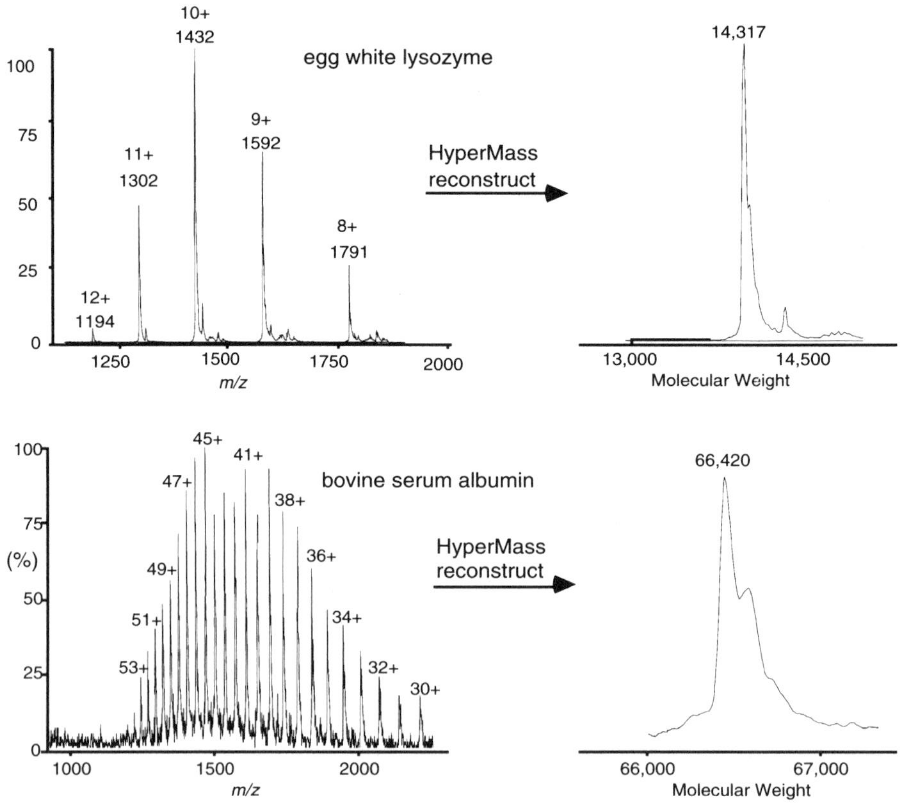

Figure 4.15 Electrospray mass spectrum and calculated molecular weight spectrum of egg white lysozyme and bovine serum albumin (BSA).

Electrospray can also be applied to glycoprotein analysis. However, in general it is not the method of choice. Direct analysis of glycoproteins by electrospray provides only limited information, and requires very high purity because excessively heterogeneous compounds produce complicated spectra, often making interpretation difficult or impossible. In addition, sample heterogeneity will often reduce instrument sensitivity. Figure 4.16 does show that electrospray is capable of providing analytically useful information for macrosized glycoproteins. Interestingly, the glycosylation in this study did not appear to seriously impair the effectiveness of the electrospray ionization, although it is expected that multiple charging on

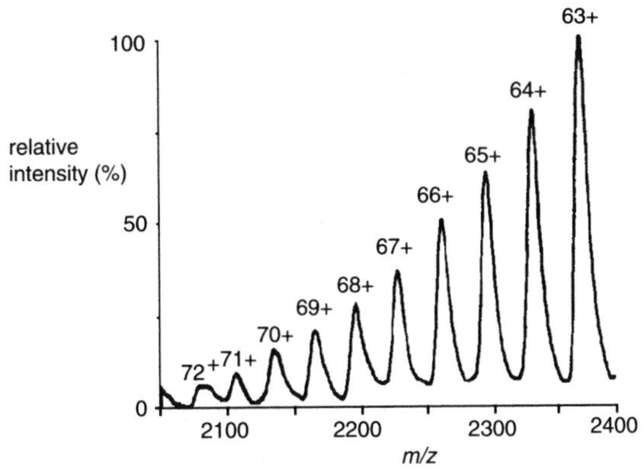

Figure 4.16 Electrospray mass spectrum of a large glycoprotein IgG-class murine monoclonal antibody, anti-(human α1-acid glycoprotein). Averaged molecular weight = 149,599 ± 12 Da. Sample consumption was 99 pmol. Adapted from Feng and Konishi (1992).

carbohydrates will not be as extensive as on peptides. MALDI–MS is still preferable for the analysis of glycoproteins.

The electrospray ionization process described in Chapter 1 induces ion formation from small droplets. Once the ions are formed, they are subject to collisions upon entering the mass analyzer. These collisions can decluster aggregates, induce fragmentation, and change the charge states by removing cations. The energy with which the ions enter the mass analyzer through the orifice can determine the amount of fragmentation that will take place. This energy can be adjusted by varying the electrospray declustering potential. In relation to peptide/protein analysis the declustering potential allows for peptides to be fragmented, thus allowing for sequencing information to be obtained. It also allows for different charge states to be observed. And by lowering the potential, electrospray becomes more conducive to the observation of noncovalent interactions.

A diagram of the orifice is shown in Figure 4.17. The energy of the ions entering the orifice is determined by the voltage applied to the orifice. Another feature of the orifice potential is how it effects sensitivity. Peptides and proteins are often very sensitive to the potential at the orifice. If the value is too high or too low, the sensitivity will drop off. The optimum value is largely compound dependent.

Increasing the declustering potential can also cause the charge distribution of a protein to change. This is related to the protein's greater collisional energy when entering the mass analyzer. At higher declustering potentials,

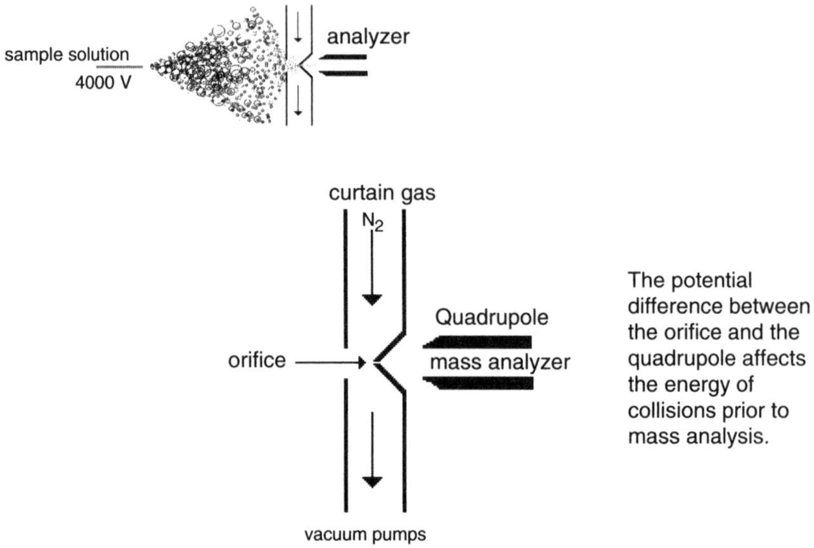

Figure 4.17 Orifice-induced collisions are controlled by the potential between the orifice and the quadrupole mass analyzer. The potential affects sensitivity, ability to observe noncovalent interactions, fragmentation, and charge state distribution.

more protons are stripped from the protein, causing a shift in the charge state distribution. Figure 4.18 demonstrates that at a high declustering potential, protons are stripped from egg white lysozyme to produce a charge distribution completely different from one observed at a lower potential.

Electrospray orifice potentials can also be manipulated to induce fragmentation and thereby obtain sequence information. Figure 4.19 is an example of how useful sequence information was obtained on a peptide simply by increasing this potential. A drawback of the application is that it can only be used on very pure samples; if the samples are not pure it may be easy to mistake an impurity for a fragment ion.

When using orifice-induced fragmentation, identifying true fragment ions versus impurities is difficult. This problem can be overcome with the use of a tandem mass spectrometer. Sample purity is not as significant an issue with a tandem mass spectrometer since the first analyzer can isolate an ion of interest even if there are several other ions (peptides) in the spectrum. Once the ion of interest is isolated, it can be guided into a collision cell and the resultant fragment ions can be mass analyzed by another analyzer. This technique is called tandem mass analysis because it

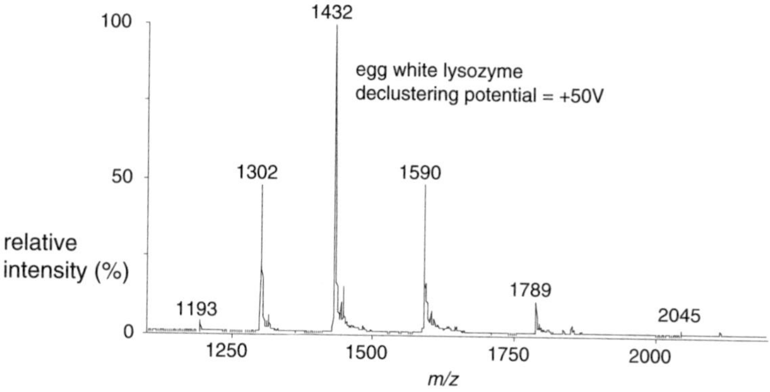

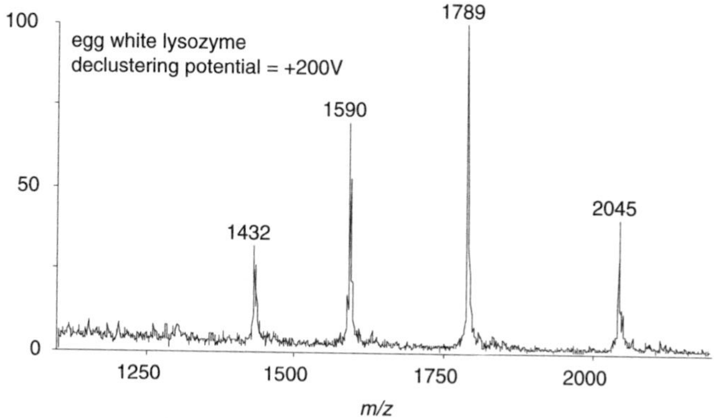

Figure 4.18 Electrospray mass spectra of egg white lysozyme at two different orifice potentials.

uses multiple analyzers (see Chapter 2). The specific applications of tandem mass spectrometry to peptides are described in the following section.

Sequencing of Peptides and Proteins

This section discusses different methods for obtaining sequence information on peptides and proteins, including tandem mass spectrometry and protein ladder sequencing.

Mass spectrometry has developed a symbiotic relationship with other technologies in determining protein structure. Having played a central role in the discovery of unique protein modifications, mass determination is also becoming more important toward obtaining total sequence information.

In general, mass spectrometry obtains structural information on peptides and proteins by collision-induced dissociation (CID), in which precur-

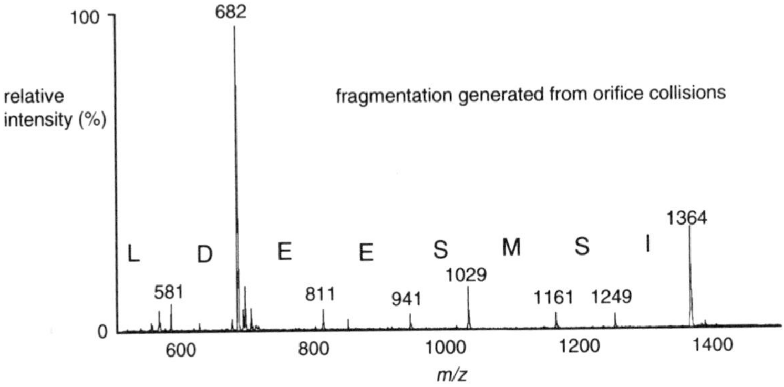

Figure 4.19 Collisions at the orifice can be useful for obtaining structural information. The collisions of this peptide, H-[ISMSEEDLLNAK]-OH, have resulted in the formation of fragment ions directly corresponding to its sequence.

sor ions are subjected to collisions and the structure of the resulting ions is characterized. As discussed in Chapter 2, CID can be accomplished with a variety of instruments, including tandem double-focusing magnetic sectors, triple-quadrupoles, Fourier transform–ion cyclotron resonance, and time-of-flight reflectron mass analyzers. The triple quadrupole combined with electrospray is currently the most common means of generating structural data, being capable of high sensitivity, producing a reasonable amount of fragmentation information, and being relatively inexpensive. MALDI with time-of-flight reflectron and Fourier transform–ion cyclotron resonance are also becoming common sources for structural information.

In order to obtain sequence information by mass spectrometry, fragments of an ion must be produced that reflect structural features of the original compound. Fortunately, most peptides are linear molecules, which allow for relatively straightforward interpretation of the fragmentation data. The process is initiated by converting some of the kinetic energy from the peptide ion into vibrational energy. This is achieved by introducing the selected ion, usually an $(M + H)^+$ or $(M + nH)^{n+}$ ion, into a collision cell where it collides with neutral Ar, Xe, or He molecules, resulting in fragmentation. The fragment ions (or daughter ions) are then monitored with a second mass spectrometer. This application of tandem mass spectrometry also allows for a heterogeneous solution of peptides to be analyzed, and then by filtering the ion of interest into the collision cell, structural information can be derived on each peptide from a complex mixture.

The strength of triple quadrupole collision-induced dissociation was demonstrated in a series of *Science* articles by Hunt and co-workers (a brief description of these studies is given in Chapter 6). In these reports MHC peptides were analyzed using liquid chromatography and tandem mass spectrometry, producing enough sequence information on femtomoles of peptide to allow for their identification. While this type of application is at the limit of the detection capabilities for most instruments and is not yet routine, it is still an impressive example of what can be accomplished with these techniques.

From the discussion thus far, it is apparent that fragmentation is a vital step in sequencing peptides and proteins. The fragment ions produced in this process can be separated into two classes. One class retains the charge on the N-terminal while cleavage is observed from the C-terminal (Figure 4.20). This fragmentation can occur at three different positions, each of which is sequence designated as types a_n, b_n, and c_n. The second class of fragment ions generated from the N-terminal retains the charge on the C-terminal, while cleavage is observed from the N-terminal. Like the first class, this fragmentation can occur at three different positions, types x_n, y_n, and z_n.

Certain limitations for complete sequence information do exist. In determining the amino acid sequence of a peptide, a complete ion series could be used except that neither leucine and isoleucine, nor lysine and glutamine can be differentiated because they have the same mass. Because a complete ion series is not usually observed, the information from both the N- and C-terminal ions must be used to help determine the entire sequence.

It may appear difficult to identify a particular ion series in a spectrum; however, there are several rules that can be used to determine the identity of peaks belonging to any ion of the backbone fragmentation (a_n, b_n, c_n, x_n, y_n, and z_n). First, backbone fragmentation can differ in *m/z* by only one of the 18 mass values associated with the amino acids, and second, fragments a_n, b_n, and y_n generally form a continuous series.

Sequence information can be obtained for peptides with molecular weights up to 2500 Da. Larger peptides reveal at least partial sequence information that will often suffice to solve a particular problem. Proteolytic digests generate peptides from which MS/MS data on different peptides can solve the structure of a relatively large protein. One useful example of LC/MS/MS involves the synthesis of HIV protease and the application of mass analysis toward the identification of impurities in this synthesis (see Chapter 6).

Even partial sequence information can be very informative. For example, sequence information has been obtained from the 2+ charge state of a rhodamine-substituted peptide. In this case, the researcher needed to know at which amino acid of the peptide the rhodamine was attached. Since the sequence of the peptide was already known, any sequence information could potentially point to where the rhodamine was bound. Once the

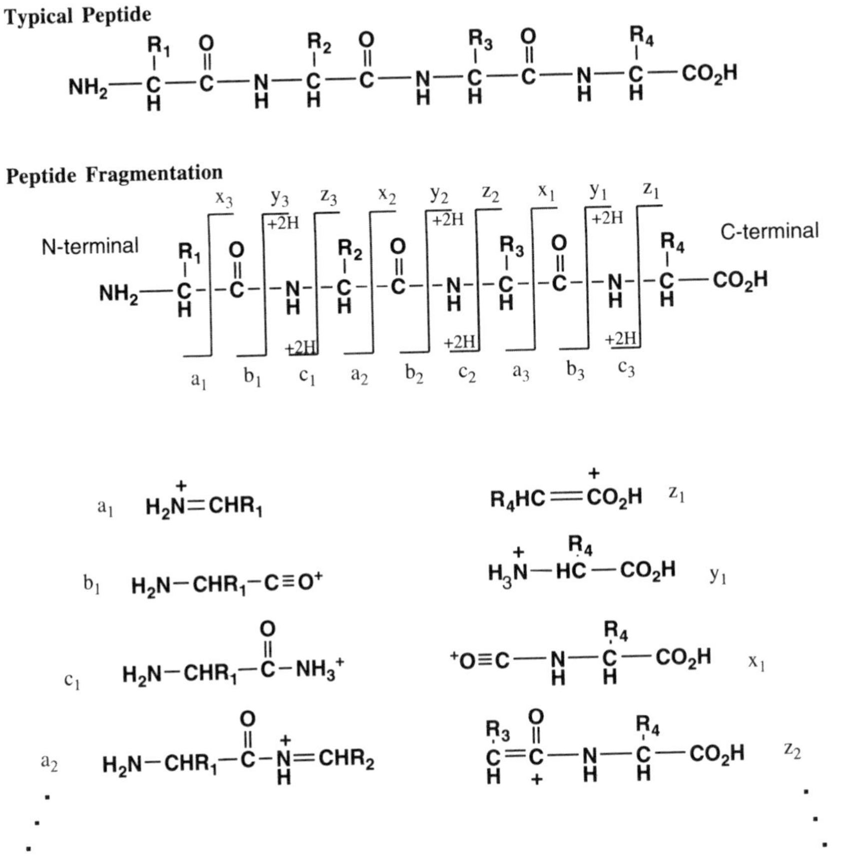

Figure 4.20 Peptide fragmentation results in the ions shown above. Collision-induced dissociation (CID) spectra often result in the dominant fragmentation at the amide bonds in the polyamide backbone, producing ions of the type B or Y.

optimal conditions on the electrospray triple-quadrupole mass spectrometer were found, the data and analysis were completed within minutes. The data shown in Figure 4.21 confirmed that the rhodamine dye was bound to cysteine.

Figure 4.22 illustrates the steps used in determining protein structure. Intially, the molecular weight of the entire protein must be obtained. Digestion is usually the next step, which generates peptide fragments. These peptides can then be analyzed both for molecular weight and sequence information to potentially provide an entire or partial sequence of the

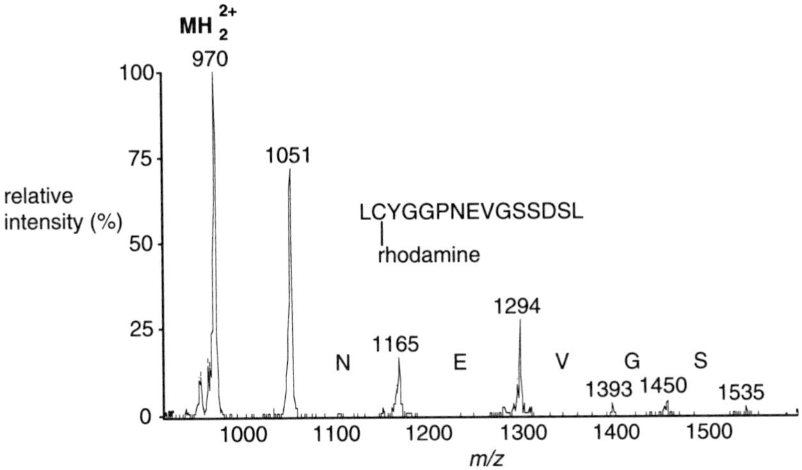

Figure 4.21 The CID data of this partial sequence and the chemistry performed on it [rhodamine-peptide]$^{2+}$ (m/z 970) enabled the researcher to identify the position of rhodamine. Rhodamine peptide courtesy Professor Klaus Hahn, The Scripps Research Institute.

protein. However, it is rare that an entire protein would be sequenced by mass spectrometry. Instead, mass spectrometry is used in conjunction with some of the more traditional sequencing techniques. It is also becoming more common to use mass spectrometry in conjunction with existing protein databases to perform peptide and protein mapping.

Peptide and protein mapping is a technique that uses mass data to obtain information on a protein. Mapping produces a mass spectrum of a peptide mixture by digesting a protein sample with a proteolytic enzyme and then analyzing the resulting digest mixture by mass spectrometry. The mass data from this spectrum can provide a match to the protein of high specificity, such that it is possible to identify the protein from this information alone. Once mass spectral data on the peptides are obtained, they are compared with a database of peptide mass values. These database values are calculated by using the enyzme cleavage rules to a collection of sequence data, such as SwissProt. The closest match to the unknown protein can then be identified. If the database does not provide a match for the unknown protein, then a search can be performed to identify entries that exhibit similar sequence homology, possibly representing equivalent proteins from related species.

Because of the high sensitivity offered by MALDI and electrospray ionization, these are the ionization techniques of choice for peptide and

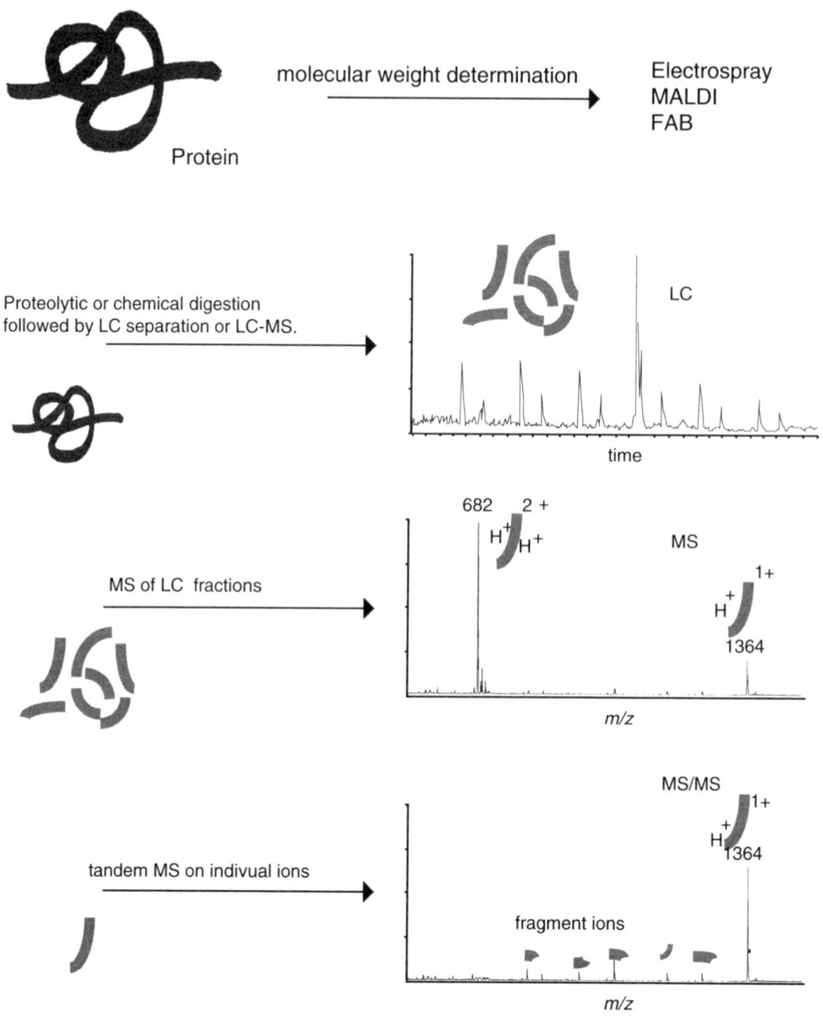

Figure 4.22 Protein characterization by mass spectrometry.

protein mapping. Peptide mapping can be performed by compiling the mass data obtained for the individual peaks in liquid chromatography. Alternatively, MALDI could be performed directly on the digest, thus eliminating the liquid chromatography separation step. These approaches, along with protein ladder sequencing, are fast making mass spectrometry a viable tool for protein structure determination.

Heterogeneity is one characteristic that can be qualitatively assessed by mass analysis, especially with MALDI–MS, which is unique in its ability to analyze extremely complex mixtures without extensive purification. Mixture analysis by MALDI has formed the basis for protein ladder sequencing as well as the mass analysis of proteolytic digests, allowing for extensive information to be gathered on the linear structure of a protein.

Protein ladder sequencing consists of a two-step process: (1) **ladder-generating chemistry,** the controlled generation of sequence-defining peptide fragments by wet chemistry, with each fragment differing from the next by one amino acid; and (2) **mass analysis,** a one-step MALDI mass analysis of the resulting protein sequencing ladder. Each amino acid is identified from the mass difference between successive peaks, and the position in the data set defines the sequence of the original peptide chain. Protein ladder sequencing has the potential for rapid sequencing.

The goal of rapidly and routinely obtaining full sequence data on peptides and proteins is coming closer with the development of protein ladder sequencing and databases that allow for rapid comparisons of proteolytic digests.

Summary

FAB, MALDI, and ESI are all effective methods for generating peptide and protein ions. The results are generally straightforward. In most cases, the ionization process involves the addition of protons. In addition, mass spectrometry has also demonstrated its utility in identifying chemically modified peptides and proteins, as these modifications usually lead to a predictable change in molecular weight. Partial structural identification of proteins is now achieved by chemical degradation followed by mass analysis of the decomposition products.

Peptide and Protein Analysis by FAB

The application of FAB to peptide analysis has several advantages. It typically generates accurate molecular weight information and it is a rapid technique. Since FAB has a limited mass range, routinely on the order of 5000–8000 Da, and relatively poor sensitivity (typically requiring >100 pmol), its application to protein analysis is not common.

Peptide and Protein Analysis by MALDI

MALDI is useful for peptide analysis in three ways: (1) it allows for rapid preparation and analysis, (2) it is very sensitive, and (3) it is tolerant of heterogeneous samples. MALDI mass analysis is typically accomplished with a time-of-flight analyzer which, with resolving capabilities on the order of 200–500, represents MALDI's biggest disadvantage. Typical accuracies for MALDI range from ±0.5 to 0.01% depending on the presence of an internal standard, the type of instrument being used, and the type of matrix material. Reflectron time-of-flight analyzers offer good accuracy with resolving power on the order of 1000–3000. Time-of-flight analyzers offer fair accuracy with resolving power on the order of 200–500.

In addition to sensitivitity, MALDI also has other useful features: the ability to measure compounds in the presence of salt (~1.0 m*M*) and the ability to analyze complex mixtures (like glycoproteins). For MALDI analysis of small molecules and peptides (200–1000 Da), the DHB matrix works well because it produces only a minimal amount of interference in the low-molecular-weight range. Basic matrices have also been introduced, such as the 2-amino-4-methyl-5-nitropyridine matrix, which extend the utility of MALDI to acid-sensitive peptides, proteins, or other acid-sensitive compounds.

Peptide and Protein Analysis by Electrospray

As with MALDI and FAB, the electrospray ionization of peptides and proteins involves the addition of a proton or multiple protons. Sample preparation is achieved by dissolving the sample in a protic-volatile solvent system which is relatively homogeneous and salt free (<500 μM). Electrospray is the method of choice for peptides and low-molecular-weight proteins (below 70,000) because of its high mass accuracy and because it can be easily interfaced with liquid chromatography, making is useful for online LC–MS.

Sequencing of Peptides and Proteins

Mass determination has played a central role in the discovery of unique protein modifications and is becoming more important toward obtaining total sequence information. In contrast to chemical detection techniques used to characterize proteins, mass spectrometry relies on precise measurement of mass.

In general, mass spectrometry obtains structural information on peptides and proteins by subjecting precursor ions to collisions. CID results in fragment ions that can lead to structural characterization. As discussed in Chapter 2, CID can be accomplished with a variety of instruments, including tandem double-focusing magnetic sectors, triple quadrupoles, Fourier transform–ion cyclotron resonance, and time-of-flight reflectron mass analyzers. The triple quadrupole combined with electrospray is currently the most common means of generating structural data, as it is capable of high sensitivity, produces a reasonable amount of fragmentation information, and is relatively inexpensive. Fragmentation information generated in MALDI time-of-flight reflectron and Fourier transform–ion cyclotron resonance mass spectrometers is also becoming more common. A dramatic illustration of mixture analysis is shown below with protein ladder sequencing, where a mixture of peptides/proteins is formed by performing Edman degradation. The degradation mixture is then MALDI mass analyzed allowing for sequence determination. Another application of MALDI, the analysis of peptides generated from digests of a protein, is used for peptide mapping.

References

Chait, B. T., Wang, R., Beavis, R. C., and Kent, S. B. H. (1993). *Science* **262,** 89–92.

Feng, R., and Konishi, Y. (1992). *Anal. Chem.* **64,** 2090–2095.

Fitzgerald, M. C., Parr, G. R., and Smith, L. R. (1993). Basic matrices for the matrix-assisted laser desorption/ionization mass spectrometry of proteins and oligonucleotides. *Anal. Chem.* **65,** 3204–3211.

Huberty *et al.* (1993). *Anal. Chem.* **65,** 2791–2800.

Martin de Llano, J. J., Jones, W., Schneider, K., Chait, B. T., Manning, J. M., Rodgers, G., Benjamin, L. J., and Weksler, B. (1993). *J. Biol. Chem.* **268,** 27004–27011.

Nelson, R. W., Dogruel, D., and Williams, P. (1994). *Rapid Commun. Mass Spectrom.* **8,** 627–631.

Schindler, P. A., Van Dorsselaer, A., and Falick, A. M. (1993). Analysis of hydrophobic proteins and peptides by electrospray mass spectrometry. *Anal. Biochem.* **213,** 256–263.

Vorm, O., Reopstorff, P., and Mann, M. (1994). Improved resolution and very high sensitivity in MALDI time-of-flight surfaces made by fast evaporation. *Anal. Chem.* **66,** 3281–3287.

CHAPTER 5

Carbohydrates, Oligonucleotides, and Small Molecules

This chapter describes carbohydrate, oligonucleotide, and small molecule analysis with the FAB, MALDI, and electrospray ionization techniques.

Perspective

The preceding chapters have typically discussed peptides and proteins to exemplify the capabilities of mass spectrometry. While peptides and proteins are the most straightforward and popular subjects for mass analysis, they represent only a small portion of biological material. FAB, MALDI, and electrospray also work well for carbohydrates, oligonucleotides, and most other biomolecules (Scheme 5.1) with results that match or exceed the performance observed for peptides and proteins. These results are summarized below.

Carbohydrates–Molecular weight information on carbohydrates can routinely be obtained using FAB, MALDI, and electrospray. MALDI and electrospray offer subpicomole sensitivity.

Oligonucleotides–Electrospray and MALDI have been successful in oligonucleotide analysis. Electrospray produces excellent accuracy and reproducibility for oligos up to 30 bases and as high as 90. Recent reports also provide sequence information.

Small molecules and natural and synthetic products–FAB, MALDI, and electrospray are all applicable to a wide variety of molecules with excellent sensitivity (subpicomole) achieved with electrospray and MALDI.

Structural information–Electrospray and MALDI combined with tandem mass analysis also offer a means for partial or total structure

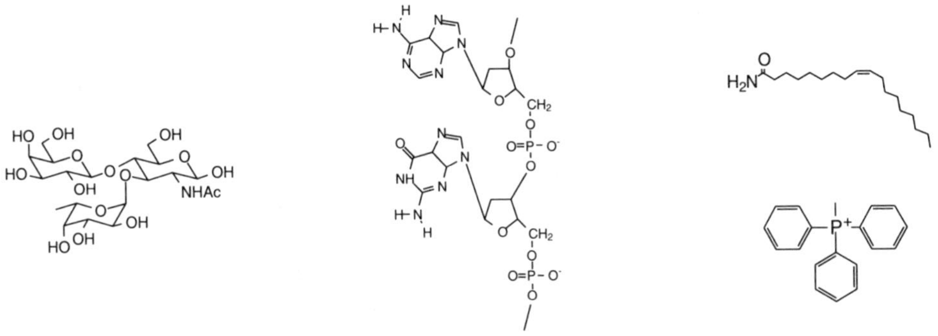

Scheme 5.1

determination for carbohydrates, oligonucleotides, and a host of synthetic and naturally derived products.

Carbohydrates, oligonucleotides, and a myriad of other biomolecules play an essential role in biological processes. Mass analysis offers sufficient sensitivity for their analysis and the possibility of partial characterization of their structures and functions.

Peptide and protein ionization, usually involving protonation, is straightforward. However, while some biomolecules will form stable protonated species, it is also common for a compound to be unstable to protonation. Their analysis, therefore, incorporates alkali cationization, deprotonation, electron ejection, and protonation. Carbohydrates, for instance, often do not form stable protonated molecular ions, while deprotonation and alkali cationization are especially effective (Chapter 2). The goal of this chapter is to provide a basic overview of the diversity of FAB, MALDI, and electrospray and how these other compounds are typically analyzed.

FAB Analysis

Even though FAB is the least sensitive and has the most limited mass range of the three ionization techniques, I have opted to discuss FAB because it is a versatile ionization tool and is still routinely used in many laboratories. At Scripps over 40,000 analyses using FAB have been performed; over 99% of the compounds were biological in origin or synthetically derived biological agents, and less than 5% of these were peptides. A list of the types of compounds that FAB has helped to characterize would be far too long; however, suffice it to say that most compounds with polar functional groups can be analyzed by FAB.

While other matrix materials (Chapter 3) are also used, nitrobenzyl alcohol (NBA) is the standard FAB matrix because it is amenable to many different compounds. In addition to matrix material there are matrix additives that are mixed with compounds that are not amenable to protonation. However, if all you need to do is protonate your compound, neat NBA with perhaps a few microliters of cosolvent should perform adequately (see Figure 5.1).

In FAB analysis there are two popular types of matrix additives—salts and cation traps. Sodium iodide or cesium iodide are used for cationization and are effective matrix additives because many compounds will not form stable protonated molecular ions, and protonation can result in significant fragmentation with virtually no observable molecular ion. Stable formation of the molecular ion can be achieved by alkali cationization (Chapter 1). Another benefit of adding alkali cations is that it will produce more reference (or matrix) ions in the spectrum. These additional peaks can be an inconvenience when interpreting the data, yet they can also be valuable for verifying the calibration of your instrumentation and as reference ions in exact mass measurements. Cesium iodide is a preferred additive in our laboratory because it provides reference ions as high as *m/z* 1400. We typically add 2 μl of a saturated CsI aqueous solution directly to the NBA matrix. Figure 5.2 illustrates the use of CsI as an additive to NBA in the analysis of a derivatized sugar. It should be noted that acidic compounds will often not respond well to the addition of salt and can lead to signal suppression. Alternatively, acid compounds work well in the negative ionization mode.

If the compounds being analyzed are very acidic and cannot be readily cationized, it may be necessary to perform FAB in the negative-ionization mode. The sensitivity of negative FAB can be improved by the addition

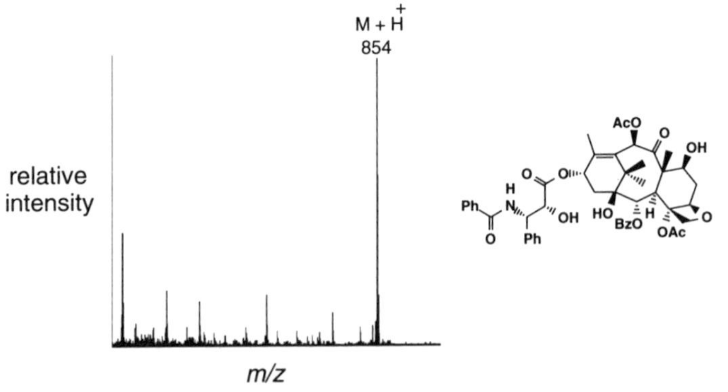

Figure 5.1 A FAB mass spectrum of synthetically derived taxol using NBA as a matrix and methanol as a cosolvent. (From Nicolaou *et al.*, 1994).

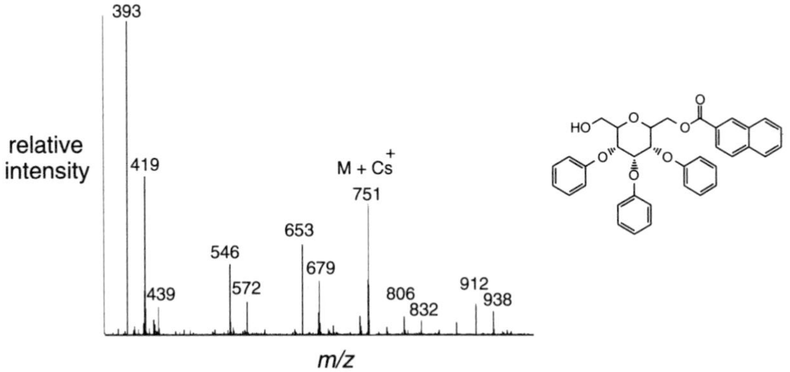

Figure 5.2 This spectrum represents how matrix ions can help generate high-accuracy data and at the same time provide for cationization. This derivatived sugar was analyzed in NBA/CsI, and all of the labeled ions (except 751) are matrix. These ions are useful for exact mass analysis and can be useful for calibration (Chapter 1).

of another matrix additive. For instance, cation traps like crown ethers and *n*-alkylnicotinium halides enhance negative ion FAB sensitivity. Figure 5.3 illustrates the utility of a matrix additive, alkylnicotinium bromide, in the analysis of coenzyme A.

MALDI Analysis

The MALDI analysis of biomolecules other than peptides and proteins is just beginning to be examined. However, its applicability appears to be

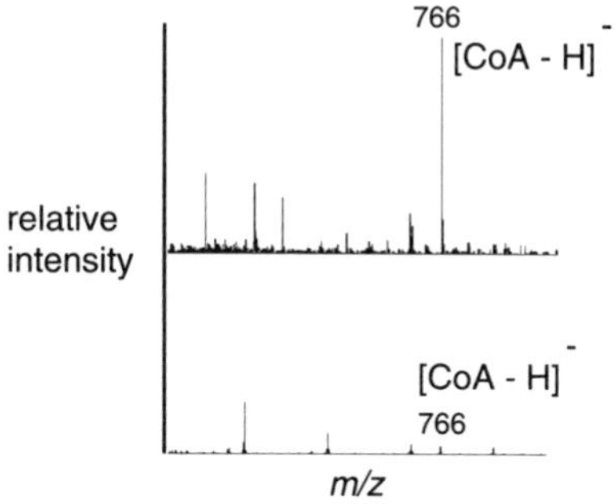

Figure 5.3 The negative ion FAB mass spectra of coenzyme A sodium salt; (top) with the addition of 1 μg of *N*-octylnicotinium bromide and (bottom) without the *N*- octylnicotinium bromide surfactant. Glycerol was used as a matrix. The intensity of the coenzyme A increased by a factor of 37 when the surfactant was added. (From Huang *et al.*, 1994).

just as diverse as FAB. This is illustrated by the number of papers that have been published on MALDI since commercial instruments became available in 1991, and by the different types of compounds that have been analyzed. Table 5.1 lists some of the compounds examined using MALDI, showing its potential for broad application.

DHB is the standard MALDI matrix (Chapter 3) used in positive ion analysis. Among its attributes is high efficiency, which translates into good sensitivity and relatively little background ionization. The lack of matrix ions in the mass spectra make DHB especially amenable to low-mass compounds, yet it is also used for high-molecular-weight species. DHB could be considered the NBA of MALDI, finding utility for a large variety of compounds in both the positive and negative ionization modes. A brief preparation protocol is described below:

Matrix: make a 10 mg/ml solution of 2,5-dihydroxybenzoic acid in 20% ethanol or water.
Dilute sample solution to a concentration of 1–10 pmol/μl with the matrix solution.
Apply 0.5–1.0 μl on the sample target.
Use a gentle stream of cold air from a blower to assist sample drying.

DHB, however, is not as effective in negative ionization as it is in positive ionization. Picolinic acid and 3-hydroxypicolinic acid are increasingly being

TABLE 5.1
Biomolecules Analyzed Using MALDI

Carbohydrates	Cyclic glucans, protected and unprotected carbohydrates
Oligodeoxynucleotides	Oligodeoxynucleotides, chemically modified oligodeoxynucleotides, polyaromatic hydrocarbon–DNA adducts, polycyclic aromatic hydrocarbon dihydrodiol epoxide DNA adducts, RNA up to 150 kDa
Natural products	Taxol, taxol derivatives, dynorphin A, rhodopsin, bacteriorhodopsin
Aromatic polyester dendrimers	
Cyclic siloxanes	
Kerogens	
Porphyrins	
Metalloporphyrins	
Organoarsenic compounds	
Cationic drugs	

used for the analysis of negative ions of oligonucleotides and even proteins. The following preparation protocol is derived from Detlev Suckau, Wang and Biemann (1994), Wu *et al.* (1993), and Nordhoff *et al.* (1992).

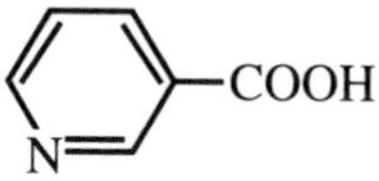

picolinic acid

Matrix: 2:1 mixture of 3-hydroxypicolinic/picolinic acid (70 mg/ml) in 50% acetonitrile.

LC-pure oligonucleotides at a concentration of 10–50 pmol/μl in water.

Prepare NH_4^+ to form ion exchange beads.

Mix 2 μl matrix and 0.5–1 μl of sample on the target, cover $\frac{1}{4}$ of the sample area with ion exchange beads, and dry using a stream of warm air.

Take *negative ion mode spectra not exceeding −25 kV,* leave the beads on the target; regions close to the beads give best signal. Positive mode spectra may also work but usually result in poor quality spectra.

As stated previously, using the proper matrix can be critical to obtaining good quality spectra (Chapter 3). Figure 5.4 illustrates the different data that can be obtained with hydroxypicolinic acid and DHB in oligonucleotide analysis. The value of the matrix selection has been demonstrated in the

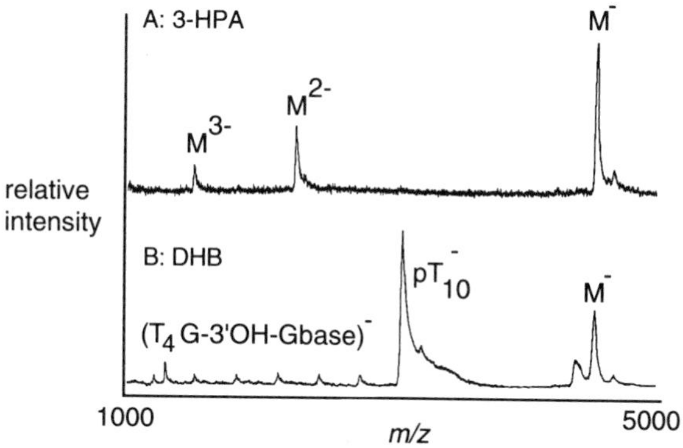

Figure 5.4 Negative ion MALDI–MS of d(TTTTGTTTTTTTTTT) generated with 355 nm radiation using the matrices 3-hydroxypicolinic acid (A) and 2,5-dihydroxybenzoic acid (B). No fragmentation is observed in A. The major fragments in B arise from phosphate backbone cleavage. (Adapted from Zhu *et al.*, 1995).

analysis of small cationic prodrugs from cell extracts (Figure 5.5) in which a prodrug chemical reaction was monitored intracellularly. These experiments were performed in the presence of cellular debris. DHB offered the lowest background and the best sensitivity over the other matrices that were used.

The study illustrated in Figure 5.5 is one of only a few that have been performed with MALDI on small molecules. Its utility to biopolymers such as DNA is gaining more recognition; two groups have demonstrated that MALDI is useful for DNA sequencing. One approach is similar to protein ladder sequencing and employs the time dependence of exonucleases to generate a sequence ladder. MALDI–MS has also shown utility in the analysis of DNA fragments generated in conventional sangor dideoxy sequencing reactions. These experiments have evaluated MALDI–MS for oligonucleotide analysis in terms of resolution, sensitivity, multiple charging, and adduct formation, suggesting that MALDI–MS may have potential as a DNA sequence analysis tool (Chapter 6). In Figure 5.6 the time-dependent exonuclease sequencing is demonstrated, where the MALDI mass analysis of the partial digestion of an oligonucleotide led to structural elucidation.

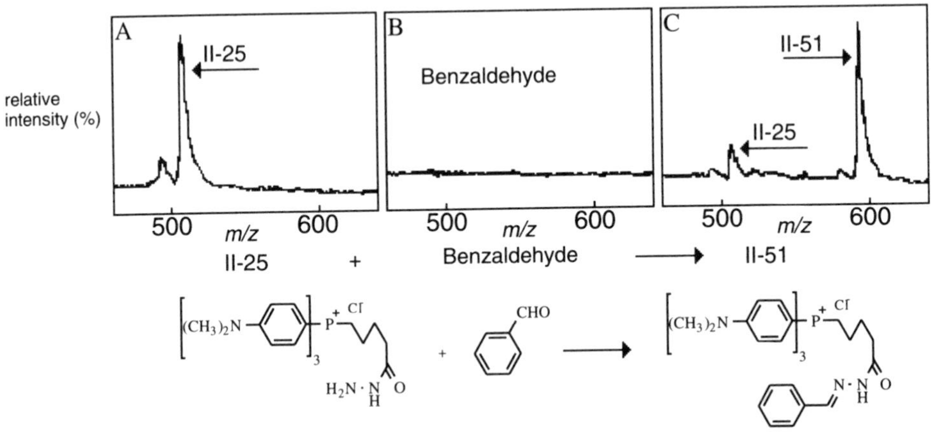

Figure 5.5 An example of monitoring intracell drug interactions using MALDI–MS. The different panels represent analysis in the presence of matrix and cellular material. The first set of cells was exposed to the hydrazone drug (II-25), the second set of cells was exposed only to benzaldehyde, and the third set of cells was exposed to benzaldehyde and a hydrazone (II-25). The interesting feature in C is that MALDI–MS was used to examine the intracell reaction products, in this case the hydrazine (II-51). (From Rideout *et al.*, 1993).

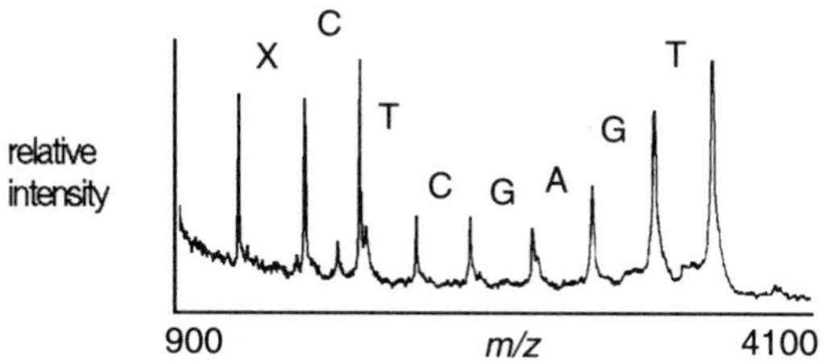

Figure 5.6 Negative ion MALDI on the partial sequence on the oligonucleotide 5′-d(GCTTXCTCGAGT) carrying a modified nucleoside in the 5 position (X = 2′-*O*-methyl adenosine). (Adapted from Pieles *et al.*, 1993).

MALDI can also analyze derivatized and underivatized carbohydrates (Figure 5.7). In contrast to peptide and protein spectra that contain protonated molecular ions, carbohydrate spectra consist primarily of $[M + Na]^+$ ions and can be obtained with subpicomole sensitivity. Analogous to the results obtained for FAB, matrix additives can also be helpful with MALDI. Likewise, negative FAB analysis of highly acidic compounds benefits from the addition of a cation trap. A different tactic has been pursued for MALDI, where acidic compounds were mixed with a basic peptide or protein, rich in arginine, to form a noncovalent complex. This additive has been useful with highly acidic polysulfated, -sulfonated, and -phosphorylated biomolecules such as heparin-derived oligosaccharides, oligonucleotides, suramin, and cysteic acid-containing peptides (Juhasz and Biemann, 1994).

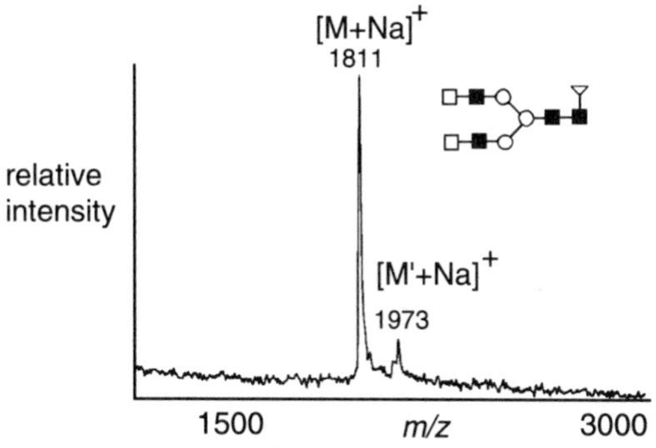

Figure 5.7 MALDI–MS of an oligosaccharide recorded with 2,5-dihydroxybenzoic acid as the matrix. The peak at *m/z* 1973 represents a second sugar containing an extra hexose residue. O, mannose; □, galactose; ■, *N*-acetylglycosamine; ▽, fucose. (Adapted from Harvey *et al.*, 1993).

In addition to producing qualitative information, MALDI has also been shown to produce quantitative results from carbohydrates, peptides, proteins, small molecules, and drugs. MALDI–MS has an inherent problem in obtaining quantitative information simply as a function of signal intensity; the nonuniform nature of the matrix/analyte solid mixture and variations in detector response often result in a nonreproducible signal response. When MALDI–MS is employed for qualitative work this is not important, as the laser fluence is simply adjusted for each sample to provide an acceptable signal. In order for MALDI to be used for quantitative analysis, it is essential to compensate for this variation. The incorporation of an internal standard allows for controlled shot-to-shot and sample-to-sample variability. A prerequisite for a suitable internal standard is that it should mimic the behavior of the analyte. In most of the studies referenced, this has been accomplished by the addition of structural analogues or stable isotope-labeled analogues as internal standards. The initial quantitative studies confirm that the use of an internal standard, such as a structural or stable isotope-labeled analogue, is necessary. Figure 5.8 illustrates the application

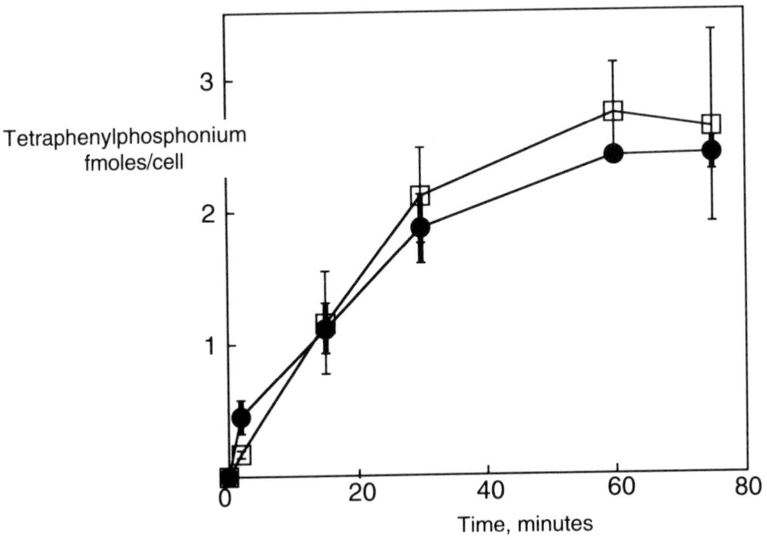

Figure 5.8 MALDI–MS data on the uptake of tetraphenylphosphonium (TPP) by carcinoma cells as a function of exposure time, as observed for unlabeled TPP with MALDI–MS (●) and for tritiated TPP with scintillation counting (□). Note the significantly smaller error bars (representing 95% confidence limits) observed for MALDI–MS. In these experiments methyl-triphenylphosphonium was used as an internal standard. (Adapted from Rideout *et al.*, 1993).

of MALDI–MS to quantifying cell–drug uptake where methyl triphenylphosphonium was used as an internal standard because of its structural similarity to tetraphenylphosphonium.

The appeal of MALDI is its sensitivity to a wide variety of compounds from the low femtomole to the picomole level. The utility of MALDI–MS for heterogeneous samples is making it even more attractive for biological samples.

Electrospray Ionization

Electrospray is the most diverse of the ionization techniques. The utility of electrospray lies in its ability to produce singly or multiply charged gaseous ions directly from an aqueous or aqueous/organic solvent system. It can generate ions from almost every type of chargeable compound whether it is of low or high mass, positively or negatively charged. Also, since there is no matrix with ESI there is little background interference. The focus of sample preparation is usually on making the sample as homogeneous and salt free as possible to maximize sensitivity.

The ability of ESI–MS to directly analyze compounds from aqueous or aqueous/organic solutions has established the technique as a convenient mass detector for high-performance liquid chromatography (HPLC). ESI also allows for mass spectrometry analysis at relatively high LC flow rates (1.0 ml/min) and high mass accuracy ($\pm$0.01%). LC–MS with ESI has been used primarily to analyze peptides and proteins, however, LC–MS performs equally as well on a wide variety of compounds, whether they are positively or negatively (e.g., oligonucleotides) charged. One such example (Figure 5.9) on the analysis of cerebral spinal fluid demonstrated the presence of the anesthetic ketamine. Ketamine is a small molecule, yet the ions were produced with relatively little to no background interference. Whether the analysis includes small or large molecules, electrospray with LC–MS is becoming more important in the analysis of biological material.

Many biological problems are now being addressed at the molecular level. Therefore the ability to characterize a compound or compounds from biological media has taken on additional importance. Beyond characterizing compounds by molecular weight, it is important to gain structural information via fragmentation. Most chemists are familiar with the fragmentation observed with electron ionization mass spectrometry. However, because ESI often produces only a minimal amount, it has been necessary to induce fragmentation on the ions formed by ESI through tandem mass spectrometry (Chapter 2). One example of the usefulness of ESI with tandem mass analysis was demonstrated at The Scripps Research Institute. They utilized electrospray's capabilities to investigate sleep and the compounds associ-

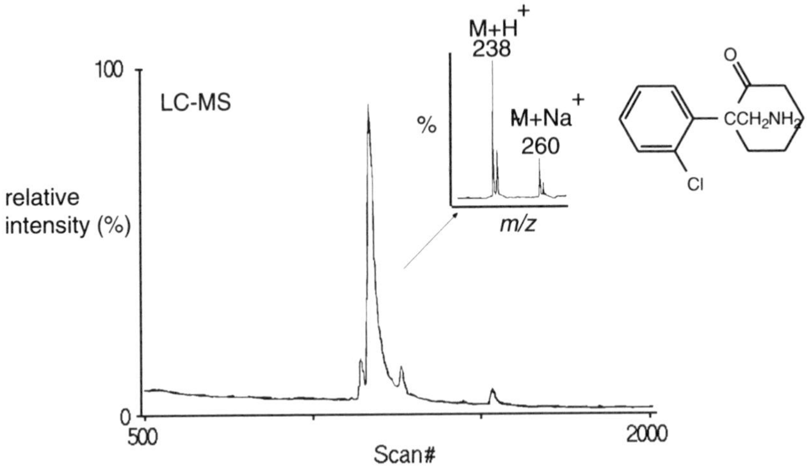

Figure 5.9 LC–MS of cerebral spinal fluid. The peak at *m/z* 238 represents the anesthetic drug ketamine. Note the isotopic pattern characteristic of the presence of chlorine.

ated with this event, extensively using LC–MS, MS^2, and MS^3. In these studies, two novel long-chain bases (MW < 350 Da) have been isolated and identified (Chapter 6).

Two additional examples of electrospray mass spectral results are displayed in Figure 5.10. The positive and negative ionization modes were used for *n*-methyl adenosine and the prostaglandin, arachidonic acid, respectively.

Oligonucleotides are another class of biomolecules that require characterization. For instance, oligonucleotides are routinely assembled using automated solid-phase DNA synthesis. This technique is normally highly efficient, having stepwise coupling yields of nucleoside phosphoramidite monomers of 99%. However, even with such efficient coupling reactions the theoretical yield for a 20mer is only 82%. Errors in synthesis can also occur, which result in a further decrease in the overall yield. Furthermore, low-molecular-weight impurities can also arise through the initiation of chain synthesis at reactive sites on the solid support or from failure sequences that are capped, blocking further chain elongation. Synthetic oligonucleotides therefore need to be separated from a large number of side products. Sometimes this can be difficult, since the impurities frequently coelute on liquid chromatography, making their detection and removal troublesome. Purification and characterization methods for oligonucleotides include HPLC, PAGE, capillary electrophoresis, and dimethoxytrityl oligonucleotide purification cartridges. These techniques are very useful;

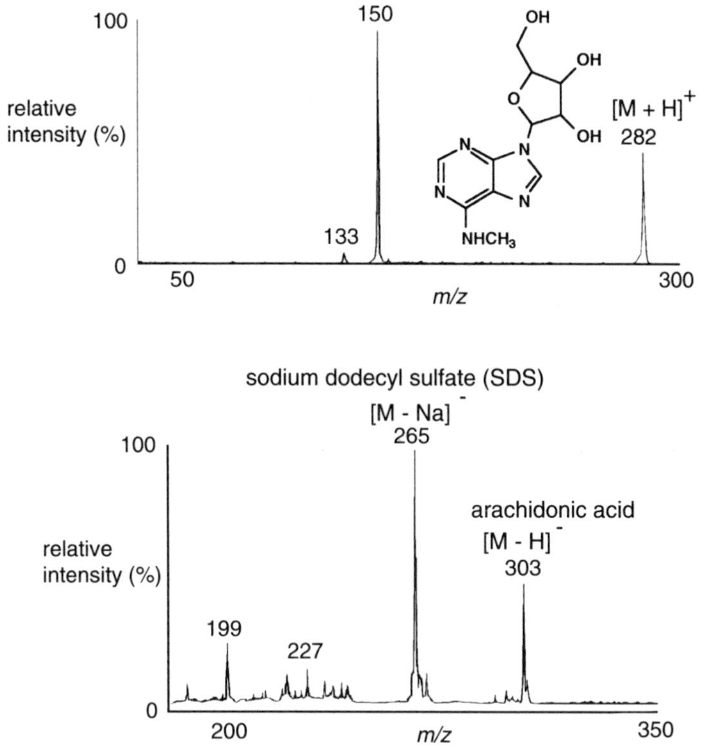

Figure 5.10 Positive ion electrospray MS/MS of *N*-methyl adenosine (top) and the negative ion electrospray mass spectrometry (bottom) of arachidonic acid in the presence of the surfactant dodecyl sulfate at *m/z* 265.

however, they cannot be used to unequivocally identify oligonucleotide composition or impurities.

Electrospray also allows for the analysis of oligonucleotides with accuracy on the order of 0.01%. As shown in Figure 5.11, the oligos provide excellent results when run in the negative-ionization mode. The typical mass range for oligonucleotides is up to 25,000 Da or 80 bases; however, oligos with up to 130 bases have also been observed. The analysis of oligos entails a rate-limiting step in which the samples must undergo rigorous desalting prior to analysis. Without this step a significant number of cation adducts, usually of sodium and potassium, will skew the mass to higher values and result in an overall reduction in sensitivity.

The use of mass spectrometry in the analysis of oligonucleotides and the more recent application of liquid chromatography and electrospray mass spectrometry to oligonucleotides are promising approaches that can

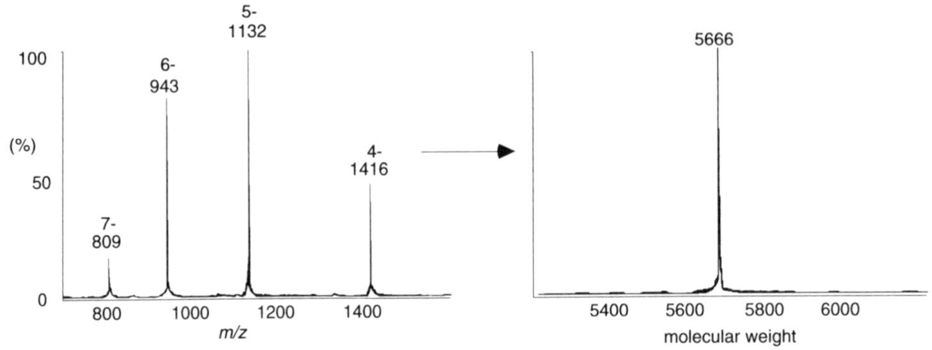

Figure 5.11 Negative ion electrospray mass spectrum of 3′-CTCGATACGCTCCGC-TAC-5′ at a concentration of 7.1 μM or 0.04 mg/ml. The reconstructed molecular weight spectrum on the right was computer generated from the mass spectral data on the left.

provide enough information to verify base composition. In addition to separating individual oligonucleotides, LC–MS has an advantage over direct mass spectrometry analysis in its ability to remove salts that can form cation adducts with the highly polar oligonucleotides and can diminish signal intensity.

The development of rapid, simple, and accurate methods for characterizing and insuring the integrity of oligonucleotides is especially important for the development of therapeutic and diagnostic products. Increasing interest in modified synthetic oligonucleotides that perform poorly on reverse-phase HPLC and whose structural integrity must be verified has heightened the need for an analytical method that can provide information on purity and structure. In a study shown in Figure 5.12, three types of oligodeoxyribonucleotides have been analyzed, including oligodeoxyribonucleotides with a phosphodiester backbone, a phosphorothioate backbone, and a methylphosphonate backbone. Each could be separated and identified even when they differed by as little as one base. The ability to detect contaminating species (at a 1% molar ratio) that coelute on reverse-phase HPLC has also been demonstrated.

Electrospray is also useful for carbohydrate analysis. It is not unusual to observe carbohydrates in a variety of different ion forms. The electrospray mass spectra of sialyl Lewis X, a cell-surface carbohydrate (Figure 5.13), is representative of what one can expect when analyzing a carbohydrate, the $[M + Na]^+$ cation. However, since this species is an acid, it was also possible to observe the same molecule in the negative ion mode.

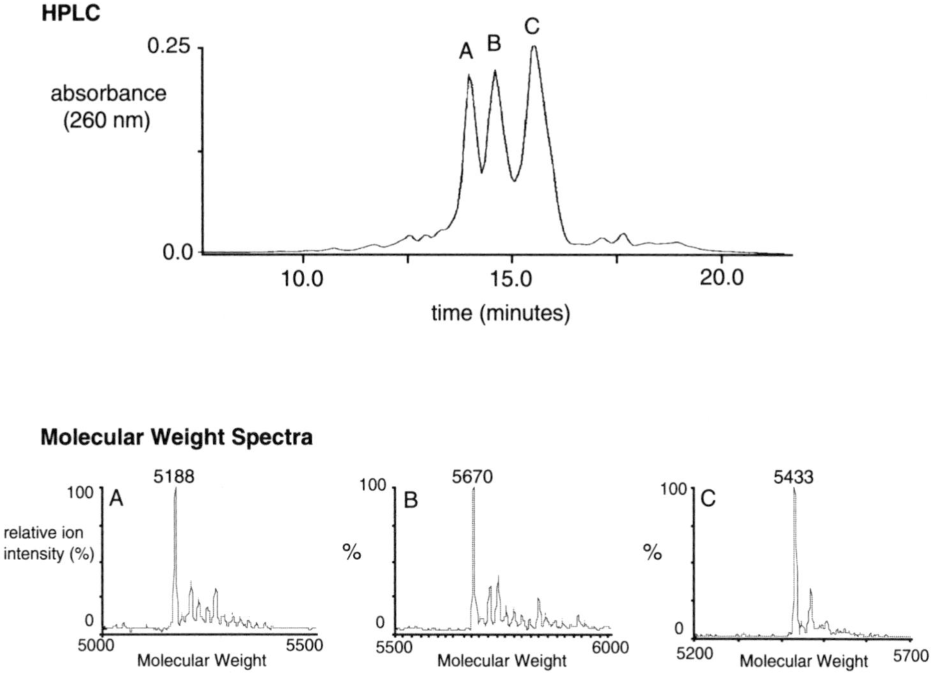

Figure 5.12 LC–MS data acquired from a single run of a mixture of three oligonucleotides. The HPLC–MS data show the reconstructed molecular weight spectra corresponding to a 17mer, and two 18mer oligonucleotides, respectively. Adapted from Bothner *et al.*

In many ways results with carbohydrates demonstrate the strength of the electrospray technique. Carbohydrates are typically labile compounds that easily fragment and can easily decompose when exposed to heat. Yet ESI has little trouble in generating these molecules in the gas phase and analyzing them in both the negative or positive ionization mode. In addition, carbohydrates will readily fragment allowing for some structural determination.

Summary

FAB Analysis

FAB is the least sensitive and has the most limited mass range of the three ionization techniques, yet it is a versatile ionization tool and is still routinely used in many laboratories. NBA is the standard FAB matrix

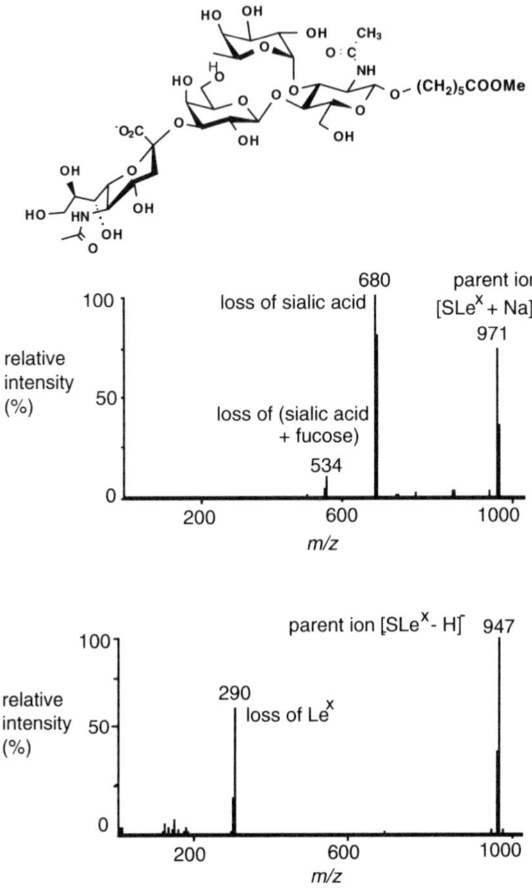

Figure 5.13 Sialyl Lewis X (top), electrospray tandem mass analysis (MS^2) of sialyl Lewis X ions, $[SLe^x + Na]^+$ and $[SLe^x - H]^-$.

because it is amenable to many different compounds. Two popular types of matrix additives are salts and cation traps. Sodium iodide or cesium iodide are used for cationization. Cation traps, crown ethers, and *n*-alkylnicotinium halides enhance negative ion FAB sensitivity.

MALDI Analysis

The MALDI analysis of biomolecules other than peptides and proteins is just beginning to be examined. However, its applicability appears to be just as diverse as that of FAB. DHB is the standard MALDI positive ion matrix; however, there are many more. Picolinic acid and 3-hydroxypicolinic acid are standard negative ion matrices. MALDI's utility for biopolymers such as DNA is gaining more recognition; its capacity to analyze derivatized and underivatized carbohydrates is also being recognized. Basic peptide additives have also been useful in negative ion analysis of highly acidic polysulfated, -sulfonated, and -phosphorylated biomolecules. MALDI has also been shown to produce quantitative results through the incorporation of an internal standard. The appeal of MALDI is its sensitivity to a wide variety of homogeneous and heterogeneous compounds from the low femtomole to picomole level.

Electrospray Ionization

Electrospray is the most diverse of the ionization techniques. The ability of ESI–MS to directly analyze compounds from aqueous or aqueous/organic solutions has established the technique as a convenient mass detector for HPLC. ESI also allows for mass spectrometry analysis at relatively high LC flow rates (1.0 ml/min) and high mass accuracy ($\pm 0.01\%$) and with low-molecular weight compounds. Oligonucleotides are another class of biomolecules that can be characterized directly or through LC–MS with a mass range of up to 25,000 Da or 80 bases.The analysis of oligos entails a rate-limiting step in which the samples must undergo rigorous desalting prior to analysis. Electrospray is also being applied to carbohydrate analysis.

References

Bothner, B., Chatman, K. S., Sarkisian, M., and Siuzdak, G., submitted.

Harvey, D. J. (1993). *Rapid Commun. Mass Spectrom.* **7,** 614–619.

Huang, Z. H., Shyong, B. J., Gage, D. A., Noon, K. R., and Allison, J. (1994). *J. Am. Soc. Mass Spectrom.* **5,** 935–948.

Juhasz, P., and Biemann, K. (1994). *Proc. Natl. Acad. Sci. USA* **91**(10), 4333–4337.

Nicolaou, K. C., Yang, Z., Liu, J. J., Ueno, H., Nantermet, P. G., Guy, R. K., Claiborne, C. F., Renaud, J., Couladouros, E. A., Paulvannan, K., *et al.* (1994). *Nature* **367,** 630–634.

Nordhoff, E., Ingendoh, A., Cramer, R., Overberg, A., Stahl, B., Karas, M., Hillenkamp, F., and Crain, P. F. (1992). *Rapid Commun. Mass Spectrom.* **6**(12), 771–776.

Pieles, U., Zürcher, W., Scär, M., and Moser, H. E. (1993). *Nucleic Acids Res.* **21,** 3191–3196.

Rideout, D., Bustamante, A., and Siuzdak, G. (1993). *Proc. Natl. Acad. Sci. USA* **90**(21), 10226–10229.

Wang, B., and Biemann, K. (1994). *Anal. Chem.* **66,** 1918–1924.

Wu, K. J., Steding, A., and Becker, C. H. (1993). *Rapid Commun. Mass Spectrom.* **7,** 142–146.

Zhu, L., Fitzgerald, M. C., Parr, G. R., Nelson, C. M., and Smith, L. M. (1995). *J. Am. Chem. Soc.* **117,** 6048–6056.

CHAPTER 6

Specific Applications

Don't let your bias keep you from doing a good experiment—Richard A. Lerner and Stephen B. H. Kent

Throughout the first five chapters the capabilities which mass spectrometry has to offer were discussed. I would like to change course from surveying the capabilities to how it is being applied to real biochemical issues. In the foregoing a cursory look at what is possible with the new generation of mass spectrometers, I have not delved too deeply into any one subject. The following final chapter is less nebulous. Presented here are condensed versions of six studies, each section being the summary of one or more papers describing unique applications of mass spectrometry.

Chemical Characterization of a Family of Brain Lipids with Sleep–Inducing Potential

The concept that there are endogeneous compounds associated with sleep has a long and complicated history, during which a large variety of substances have been proposed. Applying the sensitivity of modern analytical techniques to the study of sleep, a fresh, purely chemical approach was taken in attempting to identify molecules of the central nervous system.

Specifically, the sleep–wake states of feline subjects were examined in the following study. Cerebral spinal fluid (CSF) was removed from the feline subjects and then, using liquid chromatography, its components were separated. Electrospray tandem mass spectrometry, gas chromatography–mass spectrometry (GC–MS), and thin-layer chromatography (TLC) were also employed in the analysis of the CSF, along with infrared spectroscopy (IR), nuclear magnetic resonance spectroscopy (NMR), and chemical degradation procedures. The goal was to identify new molecules associated with the sleep–wake cycle.

CSF analysis began with preparative liquid chromatography fraction collection. These experiments produced UV data on each fraction to determine any differences between the felines at various points in their sleep cycle. An absorbance was found to be particularly prominent in the CSF of cats that were kept awake for an extended period of time (18 hr).

Even though the compound associated with this absorbance was only present in small amounts, partial characterization was initially obtained by performing exact mass measurements and tandem mass analysis. Using an API III Perkin Elmer SCIEX triple-quadrupole mass spectrometer, electrospray mass analysis on the fractions associated with the differences in the chromatogram produced a significant ion at *m/z* 282. That was determined to be the MH^+ ion, and an exact mass determination on the unknown compound by FAB (Fisons/VG ZAB-VSE) was consistent with the molecular formula $C_{18}H_{35}NO$.

CID was used to perform MS^2 and MS^3 experiments on the ion at *m/z* 282. Tandem mass analysis (Figure 6.1) at *m/z* 282 revealed a distinct fragmentation pattern in the low-molecular-mass range, consistent with other long-chain alkanes. Neutral losses of 17 and 35 Da from the parent ion indicated a loss of ammonia followed by water. Performing additional MS^3 experiments on the daughter ions at *m/z* 265 and 247 revealed that the daughter ion at 265 fragmented to form the granddaughter ion at 247. This suggested that the ion at 247 was the result of sequential losses (loss of 17 Da {NH_3} followed by 18 Da {H_2O}), as opposed to a neutral loss independent of the daughter ion at 265. Additional deuterium exchange experiments were consistent with at least two protons on this molecule being exchangeable.

On the basis of these experiments, compounds that best corresponded to the data were synthesized and the resullts obtained on these compounds

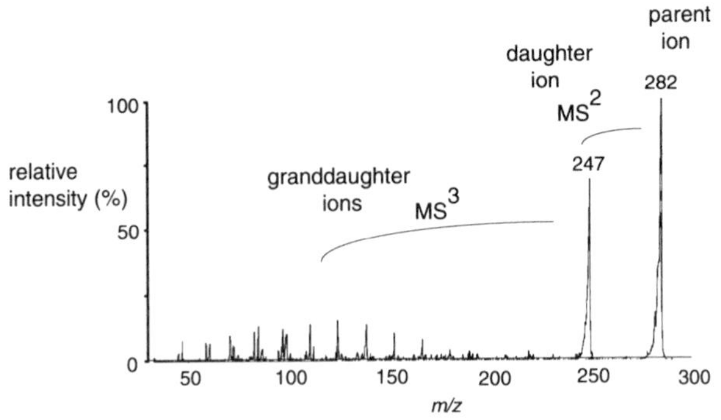

Figure 6.1 Mass spectral data (MS, MS^2, and MS_3) obtained on sleep-inducing substance.

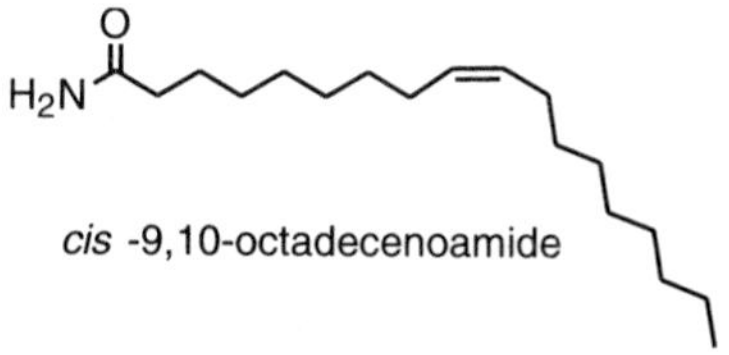

Figure 6.2 Sleep-inducing substance isolated from spinal fluid.

eventually led to the speculation that the unknown was *cis*-9,10-octadecenoamide (Figure 6.2). Additional natural compound was obtained for NMR, IR, and chemical degradation studies.

Chemical degradation techniques were first employed on synthetic fatty acid amides, identifying ozonolysis as conducive to the analysis of these agents. GC–MS analysis of the ozonolysis reaction mixture derived from the natural lipid revealed nonyl aldehyde as the only C-terminal aldehyde present. Nonyl aldehyde corresponds to an olefin positioned seven methylenes away from the terminal methyl group of the alkyl chain, which in the case of a C18 fatty acid primary amide, is the 9,10 position. The NMR and IR spectra of *cis*-9,10-octadecenoamide and the natural compound were also found to be identical.

Thus, at the conclusion of this effort, with the employment of mass spectrometry, GC, GC–MS, TLC, IR, NMR, and ozonolysis, the exact structure of the endogenous lipid, including the position and configuration of its olefin, was unambiguously determined to be *cis*-9,10-octadecenoamide.

Following its identification, synthetic *cis*-9,10-octadecenoamide was injected into rats in order to test its effect. It induced a marked, long-lasting motor quiescence and an eyes-closed, sedated behavior characteristic of normal sleep. Further studies on *cis*-9,10-octadecenoamide and other compounds observed in the CSF are currently being performed. It is interesting how, by allowing these analyses to be done, analytical technology is broadening the traditional view of lipid molecules as passive structural elements of cellular architecture and is increasing awareness of the active roles these agents play in transducing cell signals and modifying cell behavior.

Final confirmation of the unknown's identity was obtained by independent synthesis and comparison of the spectral data. The results are summarized in Table 6.1.

References

Cravatt, B. F., Prospero-Garcia, O., Siuzdak, G., Gilula, N. B., Henriksen, S. J., Boger, D. L., and Lerner, R. A. (1995). *Chemical characterization of a family of brain lipids with sleep inducing potential. Science* **268,** 1506–1509.

TABLE 6.1
Analysis of Sleep-Inducing Brain Lipid from Cats

Data type	Summary of data
Exact mass	$[C_{18}H_{35}NO + Na]^+$ Observed 304.2614 Actual 304.2616
Deuterium exchange	*m/z* 285 $[M–2H^+ + 3D^+]^+$, at least two exchangeable protons.
Isotope	Isotope pattern at *m/z* = 282 indicates that it is not a multiply charged species, nor does it have any elements with an unusual isotopic pattern (e.g. Cl or Br).
Cationization	Electrospray MS observation of *m/z* 304 $[M + Na]^+$ and *m/z* 320 $[M + K]^+$ confirms that M = 281.
MS^2 data	Fragment ion *m/z* 265 corresponding to $[MH - NH_3]^+$. Fragment ion *m/z* 247 corresponding to $[MH - NH_3 - H_2O]^+$. Lipid fragment ions.
MS^3 data	Fragmentation of *m/z* 265 ≫ *m/z* 247 (sequential loss) suggests the *m/z* 247 fragment is related to the *m/z* 265 fragment ion. Lipid fragment ions.
NMR data	Confirms lipid portion of molecule and configuration of double bond.
Ozonolysis with GC–MS	Identified the location of the double bond.
IR data	Confirms identity of unknown as *cis* isomer.

Note. Data from Cravatt *et al.* (1995) and Lerner *et al.* (1994).

Lerner, R. A., Siuzdak, G., Prospero-Garcia, O., Henriksen, S. J., Boger, D. L., and Cravatt, B. F. (1994). *Cerebrodiene: A new brain lipid isolated from sleep deprived cats. Proc. Natl. Acad. Sci. USA* **91,** 9505–9508.

Monitoring Antibody Catalysis

The following section highlights two examples using mass spectrometry to monitor antibody and enzyme catalysis. In the studies described, natural enzymes and designed catalytic antibodies were used in conjunction with electrospray mass spectrometry to study the analysis of noncovalent binding between an antibody and its antigen. Antibody–substrate reaction intermediates were also observed.

Antibodies are components of the basic defense mechanism in our immune system, their formation being an immunological response to the presence of foreign substances (antigens). In general, the interaction of an antibody with an antigen results in the noncovalent formation of an antigen/antibody complex. An antibody elicited by a specific antigen (hapten) exhibits shape and charge complementary to that hapten. This was exploited by chemists to design so-called catalytic antibodies capable of catalyzing specific chemical transformations. Catalytic antibodies are produced by introducing into mice haptens that structurally resemble a reaction's transition state species. Antibodies produced by accordingly immunized mice may, in effect, catalyze the reaction by binding and stabilizing the true metastable transition state which the hapten was meant to mimic.

In this first study, a single-chain catalytic antibody was used with a molecular weight of 26,419.9 Da. Previously, the antibody was found to bind tightly to the hapten shown in Figure 6.3. (Antibodies generally exhibit high affinities (<1 n*M*) for the haptens they were elicited with.) Our first effort was toward observing the antibody/hapten complex with electrospray mass spectrometry.

As described in Chapter 4, the electrospray ionization source allows for an ion's kinetic energy to be adjusted through the declustering potential. Declustering potentials on the order of 70 V or greater usually promote the dissociation of noncovalent complexes as well as covalent fragmentation, while lower potentials (<70 V) are conducive to the observation of noncovalent complexes (protein complexes have been analyzed at declustering potentials of 40 V). In this case, the antibody/hapten complex was observed even at declustering potentials greater than 130 V. Complete dissociation was observed at a declustering potential of 175 V where the

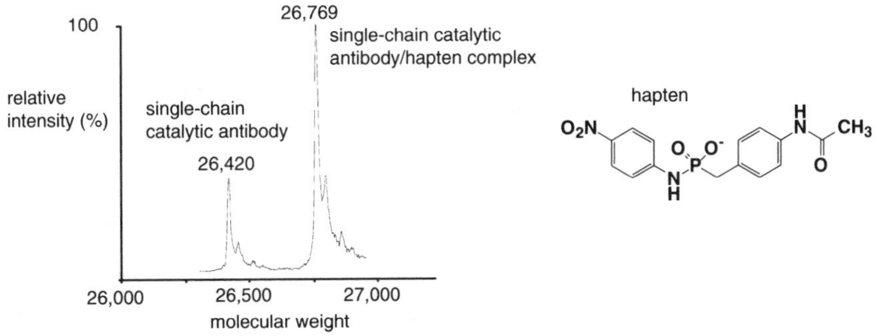

Figure 6.3 The noncovalent antibody–hapten complex reconstructed from the multiply charged states of the mass spectra. (Adapted from Siuzdak *et al.*, 1994).

antibody peak was still present, but the complex was no longer observed, consistent with its noncovalent nature. The considerably weaker binding of an inhibitor *p*-nitroaniline ($k_d = 10\ \mu M$) was indicated by the complete lack of any observable complex formation in the mass spectrum even at declustering potentials of 40 V (data not shown).

Another interesting observation in these experiments was the apparent change in the charge state distribution that occurred upon addition of the hapten. The mass spectrum of the antibody prior to hapten addition showed the presence of the charge states between 18+ and 12+. However, only two charge states, 12+ and 13+, were observed for the hapten/antibody complex at the same declustering potential (Figure 6.4). The altered charge state distribution is believed to reflect global conformational changes and proton displacement by the hapten at the binding site.

The same catalytic antibody was used to study the hydrolysis of an anilide for which this antibody was designed (Figure 6.5). Electrospray is particularly well-suited for examining covalent reaction intermediates of nucleophilic catalysis, since it can be performed at acidic pH, where hydroxide-mediated reactions are minimized. Although this technique cannot be used for the intact catalytic antibody, because of its high molecular mass (~150 kDa), it is ideally suited for examining smaller fragments such as the single-chain Fv fragment.

In this experiment the antibody was first analyzed alone, and then in the presence of substrate (Figure 6.5). Upon addition of substrate, a peak consistent with an adduct mass of 26,666 Da was observed. The observed mass increase of 247 Da is identical (within error) to the mass of the acyl functionality of the substrate. These spectra were acquired with declustering potentials of 175–250 V, under conditions where even high-affinity ($\leq nM$) noncovalent complexes are separated. Additional experiments with an antibody mutant, devoid of catalytic activity, showed no accumulation of

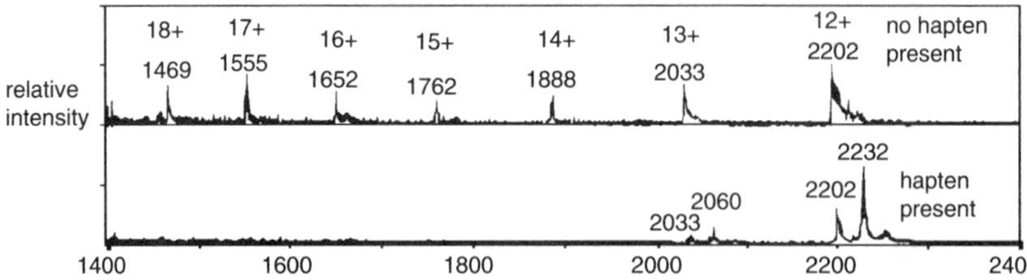

Figure 6.4 The charge state distribution of the SCA was observed under identical conditions before and after addition of the hapten. (Adapted from Siuzdak *et al.*, 1994).

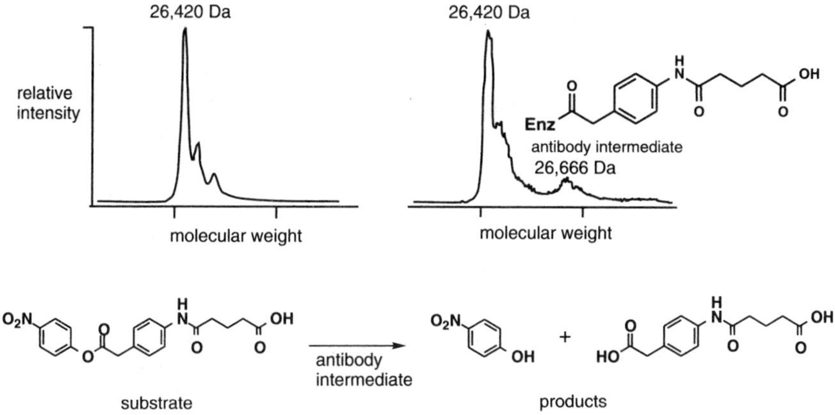

Figure 6.5 Single-chain catalytic antibody without (left) and with the substrate (right). In the presence of substrate, the formation of a covalent intermediate with the antibody can be observed. (Adapted from Krebs *et al.*, 1995).

the intermediate. These experiments provided the first direct evidence for the acyl–antibody intermediate previously proposed for this antibody.

The reaction described previously, enzyme-catalyzed ester hydrolysis, is a fundamental biochemical process. The use of fluorescence or absorption spectroscopic techniques often facilitates the kinetic study of these enzyme-catalyzed reactions. However, fluorescence or absorption spectroscopy requires that the substrate of the product have a fluorescing or absorbing chromophore. In cases in which this requirement is not fulfilled, the study of enzymes and catalytic antibodies can be severely limited. Electrospray mass spectrometry has also been used for the screening and characterization of antibodies with lipase activity since it has proven capable of analyzing proteins and small molecules with high accuracy and offers the possibility of observing compounds on the basis of their ability to be ionized, independent of the presence of chromophore.

In this study, electrospray is being used to investigate the reaction kinetics of lipid hydrolysis (Figure 6.6), along with selected ion monitoring. Initial electrospray experiments verified the identity of the observed ions as internal standard, product, and substrate. Selected ion monitoring was used to perform the calibration studies in which the product and internal standard were monitored simultaneously (Figure 6.6). Selected ion monitoring also allowed for the optimization of the experimental conditions. The electrospray kinetic measurements are currently under way.

Electrospray is a viable method for monitoring reactions that might otherwise require derivatization of the substrate. Electrospray ionization

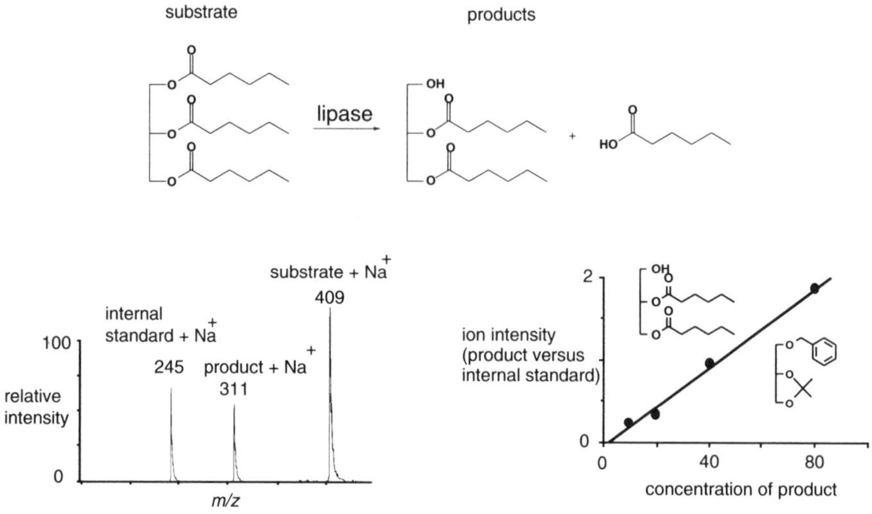

Figure 6.6 Reaction catalyzed by lipase (above), selected ion monitoring of the internal standard, product, and substrate (left), and calibration curve for relative product ion formation versus the product concentration (right).

mass spectrometry has also demonstrated its potential in the analysis of noncovalent interactions between an antibody and a hapten, and for observing covalent protein-bound intermediates in an antibody-catalyzed reaction.

References

Siuzdak, G., Krebs, J. F., Benkovic, S. J., and Dyson, H. J. (1994). *Binding of a hapten to a single-chain catalytic antibody demonstrated by electrospray mass spectrometry. J. Am. Chem. Soc.* **116,** 7937–7938.

Krebs, J. F., Siuzdak, G., Dyson, H. J., Stewart, J. D., and Benkovic, S. J. (1995). *Detection of a catalytic antibody species acylated at the active site by electrospray mass spectrometry. Biochemistry* **34,** 720–723.

Chavez, R. G., Bothner, B., Strupp, C., Hilvert, D., Siuzdak, G. (manuscript in preparation).

Protein Conformational Changes Using Deuterium Exchange and Electrospray Mass Spectrometry

Electrospray mass spectrometry has also been used to monitor protein folding. It was recognized that some proteins exhibit a distinct difference in the electrospray charge state distribution, dependent on their solution conformers. For example, two charge state distributions are shown in Figure 6.7 for a protein's native (less charged) and the denatured form (more

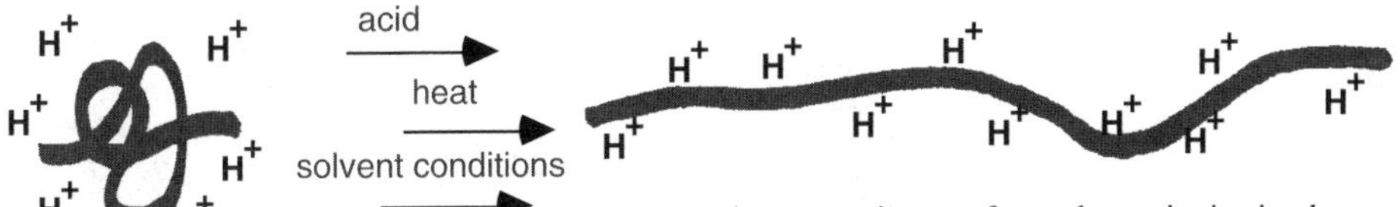

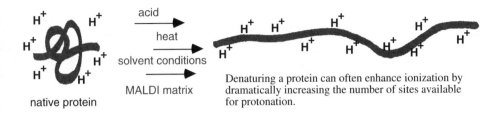

Denaturing a protein can often enhance ionization by dramatically increasing the number of sites available for protonation.

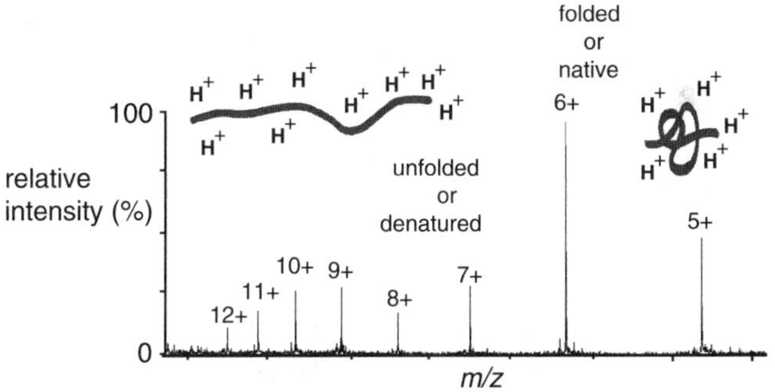

Figure 6.7 Electrospray mass analysis can be used to distinguish between native and denatured conformers of the fibronectin module. The data shown represent both the native (left of mass spectrum) and denatured conformers (right of mass spectrum) of the fibronectin module. (Adapted from Muir *et al.*, 1995).

charged). This shift in distribution is associated with the additional sites available on the denatured form for protonation. Because native proteins have fewer accessible sites for protonation, the condition of the secondary structure of the protein is reflected in the charge state distribution. For instance, a native protein like fibronectin module (shown in Figure 6.7) has a charge distribution that is maximized at the lower charge states ($\geq 7+$), while the denatured protein has a distribution that is maximized at a higher charge state (10+). The fibronectin module has a tight folded secondary structure that allows its native and denatured conformers to be separated by HPLC analysis. In this case mass spectrometry could be used as a rapid diagnostic tool to distinguish between the conformers. A similar experiment was performed on the GroEL protein, as seen in Figure 6.8.

Monitoring the charge distribution can be an effective means of distinguishing conformers; however, it provides only a limited amount of information on protein folding. Another approach toward studying protein conformational changes has been through hydrogen–deuterium exchange experiments, where the rate of exchange is affected by the folding. NMR spectroscopy, in particular, has used the hydrogen–deuterium exchange

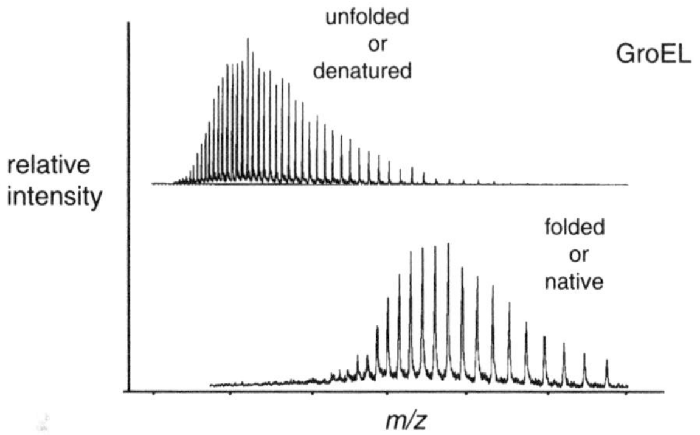

Figure 6.8 Electrospray mass spectra of the native and denatured conformers of GroEL protein. (Adapted from Robinson *et al.,* 1995).

phenomenon to monitor the average amide hydrogen exchange at individual sites. In electrospray the protein folding phenomenon can also be studied in this way; while it is not possible through ESI–MS to monitor exchange in a residue-specific manner, populations of protein molecules with distinct masses can be distinguished. Thus NMR and ESI offer complementary approaches toward studying protein and peptide intra- and interactions including monitoring protein stability and protein folding.

An example of how NMR and ESI are compatible can be shown with two populations of protein molecules. The first population has molecules that are completely exchanged and others that are not exchanged at all (Figure 6.9). The second population has 50% of the protons exchanged. Because NMR spectroscopy monitors average proton occupancy, the two populations are indistinguishable. However, ESI–MS would distinguish the two populations, as shown in Figure 6.9.

In actuality, given the large number of exchangeable protons in protein molecules and the diversity of protein motions, a range of possible masses will result. Therefore, the mass, shape, and width of the peak will be informative with respect to protein folding. The first example illustrates the ability to monitor equilibrium folding behavior of hen egg white lysozyme and how peak shape and width can change. Exchange was accomplished by diluting the deuterated protein into protonated buffer. The results (Fig. 6.10) show a steady decrease of the total mass with increasing incubation times, with a concomitant narrowing of the distribution of masses. These changes in width and shape were correlated to the dynamics of lysozyme folding.

GroEL, a protein thought to help in the folding of its substrate proteins, was also analyzed. In this case, deuterium exchange is used to distinguish

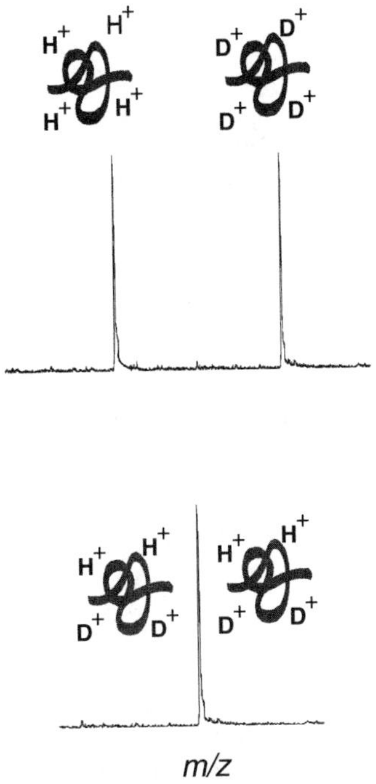

Figure 6.9 Theoretical mass spectra of two different populations of proteins.

between the numbers of protons displaced when a substrate protein is folded by GroEL bound and in comparison to globule state as is shown in Figure 6.11. In this case the uncomplexed molten globule conformer of GroEL and the GroEL bound substrate protein appear to be the same. These results suggest that the bound protein is weakly protected from exchange and is similar to globule state.

References

Muir, T. W., Williams, M. J., and Kent, S. B. H. (1995). *Detection of synthetic protein isomers and conformers by electrospray mass spectrometry. Anal. Biochem.* **224,** 100–109.

Miranker, A., Robinson, C. V., Radford, S. E., Aplin, R. T., and Dobson, C. M. (1993). *Detection of transient protein folding populations by mass spectrometry. Science* **262,** 896–900.

Robinson, C. V., Groß, M., Eyles, S. J., Ewbank, J. J., Mayhew, M., Hartl, F. U., Dobson, C. M., and Radford, S. E. (1995). *Detection of transient protein folding populations by mass spectrometry. Nature* **372,** 646–651.

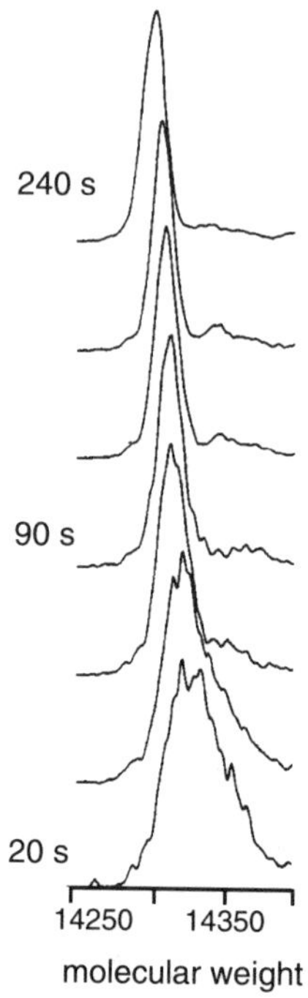

Figure 6.10 Time evolution of the ESI–MS spectra of lysozyme monitoring hydrogen–deuterium exchange. The deuterated lysozyme was dissolved in D_2O and equilibrated. After specified lengths of time, hydrogen exchange was quenched by the rapid cooling. (Adapted from Miranker *et al.*, 1993).

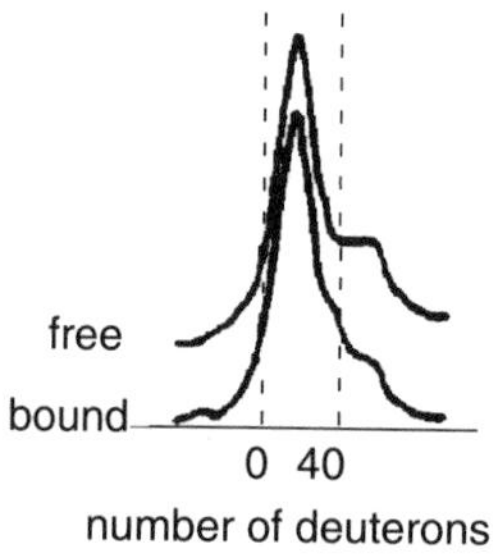

Figure 6.11 The similarity in the widths of the GroEL globule conformer peak and the peak corresponding to the GroEL-bound substrate protein suggest that the free and bound species are alike.

Characterizing Natural and Unnatural Peptides

Peptide sequencing by tandem mass spectrometry has been a useful way of characterizing peptides. In this section two separate applications are briefly described. The first involves the identification of peptides bound to MHC molecules. The technique is highly sensitive, allowing for femtomole levels of material to be characterized. The second example describes the characterization of minor by-products in peptide synthesis.

Peptides bound to MHC molecules exist in relatively small quantities. These peptides promote the recognition of cytotoxic T lymphocytes and thereby play an important role in the immune response. Identifying and characterizing these peptides has been the focus of Don Hunt and colleagues in a series of *Science* papers, their primary means of peptide characterization being liquid chromatography and electrospray ionization–tandem mass spectrometry.

The MHC molecules were purified by immunoprecipitation and the associated peptides were extracted with acid. The extracts were then prepared by microcapillary reversed-phase HPLC and then eluted directly into the electrospray mass spectrometer. In a single run, over 200 distinct ions were observed, being singly, doubly, and triply charged. A lower limit of detection was defined as 30 fmol, with a range for the detected ions up to 600 fmol. Figure 6.12 represents the total ion signal from a single chromatogram.

Some of the peptides isolated in the HPLC experiments could then be analyzed by electrospray tandem mass analysis. In these experiments, eight nonapeptides were sequenced. Figure 6.13 illustrates an example of the sequencing data for one of the peptides. This approach could eventually enhance the sensitivity and speed of the analysis of peptides. In fact, Hunt

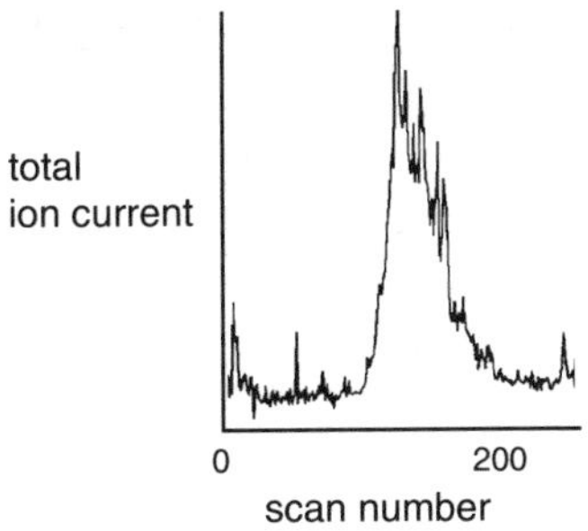

Figure 6.12 Total ion chromatogram obtained for peptides extracted from MHC molecules. (Adapted from Hunt *et al.*, 1992).

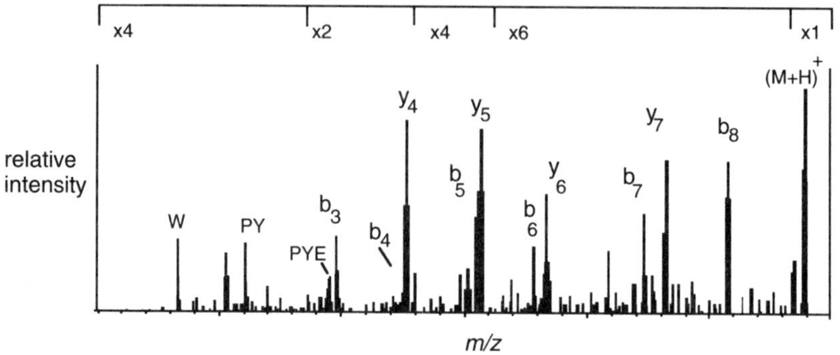

Figure 6.13 MS/MS data obtained on the peptide $(M + H)^+$ ions at *m/z*-1121 using 100–300 fmol of material derived from 2×10^8 cells. The sequence is [T Lxx W V D P Y E V]. Leucine and isoleucine are designated Lxx since they have identical mass and cannot be differentiated on the triple quadrupole mass spectrometer. (Adapted from Hunt *et al.*, 1992).

and colleagues have utilized this technique in the identification of one nonapeptide which has a high affinity for melanoma-specific cytotoxic T lymphocytes. This peptide is now being explored for its potential use in a peptide-based melanoma vaccine.

While Professor Hunt has identified natural peptide sequences, Stephen Kent and colleagues have further exploited ESI and tandem mass spectrometry toward the characterization of major and minor synthetic products in peptide synthesis. In particular, the peptide (H)-Val-Gln-Ala-Ala-Ile-Asp-Tyr-Ile-Asn-Gly-(OH) was assembled using Boc chemistry and the product was then deprotected and cleaved from the resin with HF/cresol. The crude material was purified by reverse-phase HPLC, as illustrated in Figure 6.14.

The fractions were collected and directly analyzed by mass spectrometry. One example, shown in Figure 6.15, was found to be from the cyclization of Asn73 to form a succinimide. The low-level by-products were found to be

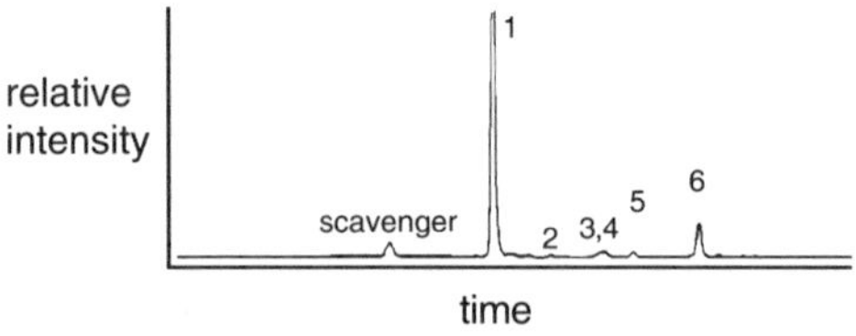

Figure 6.14 Reverse-phase liquid chromatography of crude peptide material. (Adapted from Schnölzer *et al.*, 1992).

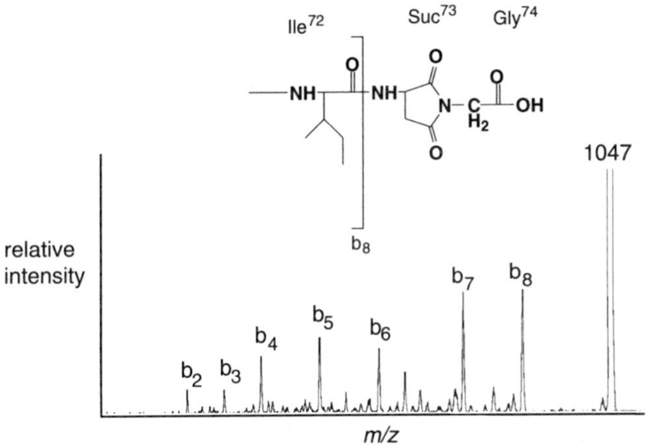

Figure 6.15 MS/MS of *m/z* 1047 parent ion from HPLC peak 2. Part of the covalent structure of the by-product arising from the cyclization of Asn73 to form a succinimide.

from incomplete deprotection or succinimide formation at Asp, succinimide formation at Asn, acylation of the Tyr, and *tert*-butylation of the decapeptide.

In the Kent laboratory the ability to perform rapid mass spectrometry and MS/MS experiments has proven essential for optimizing synthetic protocols, for characterizing peptide by-products, and for following work-up procedures.

References

Hunt, D. F., Henderson, R. A., Shabanowitz, J., Sakaguchi, K., Michel, H., Sevilir, N., Cox, A. L., Appella, E., and Englehard, V. H. (1992). *Characterization of peptides bound to the class I MHC molecule HLA-A2.1 by mass spectrometry. Science* **255,** 1261–1263.

Cox, A. L., Skipper, J., Chen, Y., Henderson, R. A., Darrow, T. L., Shabanowitz, J., Englehard, V. H., Hunt, D. F., and Slingluff, C. L., Jr. (1994). *Identification of a peptide recognized by five melanoma-specific human cytotoxic T cell lines. Science* **264,** 716–719.

Schnölzer, M., Jones, A., Alewood, P. F., and Kent, S. B. H. (1992). *Ion-spray tandem mass spectrometry in peptide synthesis: structural characterization of minor by-products in the synthesis of ACP (65–74). Anal. Biochem.* **204,** 335–343.

Protein and Oligonucleotide Sequencing by MALDI–MS

Tandem mass spectrometry has many advantages, especially when used with electrospray ionization. The previous study describes the most common way of obtaining sequence information using a triple-quadrupole mass spectrometer. Another approach has been adopted. This new approach combines the wet chemistry techniques used to degrade a sample prior to analysis with MALDI–MS.

Two important advantages of MALDI–MS are its sensitivity and ability to analyze complex polypeptide and oligonucleotide mixtures. These advantages are now being utilized to sequence biopolymers. The technique is analogous to lining up strings of beads, each having the same bead sequence (Figure 6.16A). The first string is left in its place while one bead is removed from the end of each subsequent strand (Figure 6.16B). In the next step, the first string is put aside and the rest have an additional bead removed from the same end.

This continues until a full set is obtained, with each subsequent set representing the loss of one additional bead (Figures 6.16C–6.16F). If the weight of each string of beads is measured, the difference between each subsequent string would correspond to one of the beads. This is essentially

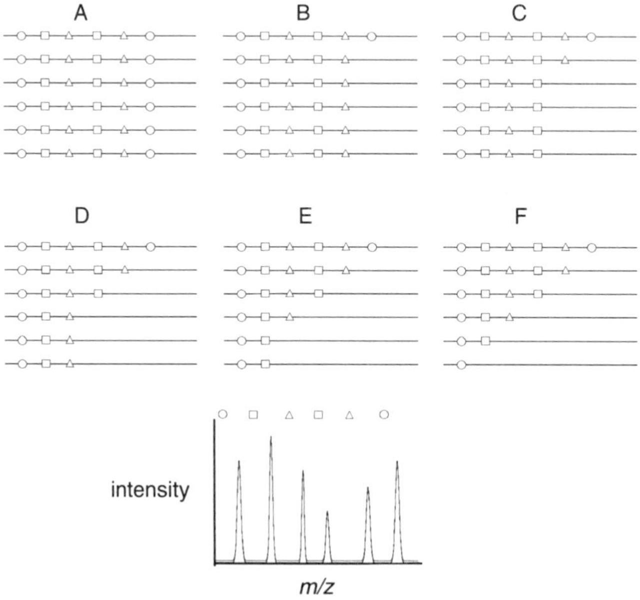

Figure 6.16 An analogy between sequencing beads on a string and protein ladder sequencing.

how MALDI produces sequence information on peptides and oligonucleotides where an amino acid or nucleotide is removed one at a time, ultimately resulting in a set that can provide this sequence information.

The removal of each amino acid in peptide and protein sequence determination is typically accomplished by Edman degradation, in which each residue is chemically removed from the amino terminal to produce sequence-defining peptide fragments. A MALDI mass spectral readout allows for the generation of the resulting protein sequencing ladder.

This method, which allows for each amino acid to be identified from the mass difference between successive peaks, provides sequence information out to more than 30 residues. Extensive sequence data can be obtained from larger proteins by enzymatic cleavage combined with protein ladder sequencing. The sensitivity of protein ladder sequencing is comparable to that of Edman degradation methods (~10 pmol), with potential for greater sensitivity.

In addition to sequencing peptides, protein ladder sequencing has the ability to directly identify post-translational modifications such as phosphorylated residues. An example of protein ladder sequencing of both phosphorylated and unphosphorylated forms of a peptide, performed by MALDI, is shown in Figure 6.17. The data allowed for the identification of the phosphorylated amino acid serine.

MALDI has also been applied to large-scale DNA sequence analysis. Analogous to the peptide/Edman degradation approach, DNA fragments are generated enzymatically in Sanger sequencing reactions, with the order of the peaks generated in MALDI corresponding to the sequence. For the separation and detection of DNA fragments that are produced in enzymatic termination reactions, conventional approaches rely on denaturing polyacrylamide gel electrophoresis coupled with radioactive or fluorescence-based detection schemes. The primary advantage of a mass spectrometry-based method would be a significant increase in the speed of the separation, detection, and data acquisition.

The feasibility of using MALDI for large-scale DNA sequencing has been investigated by Fitzgerald and colleagues. In this work, mixtures of synthetic oligonucleotides were subjected to MALDI analysis in a mock sequencing experiment. The results (shown in Figure 6.18) highlight the potential of MALDI-based DNA sequencing, as well as the considerable challenge remaining. More than an order of magnitude increase is needed in both mass range and sensitivity before the method will be competitive with existing approaches.

References

Chait, B. T., Wang, R., Beavis, R. C., and Kent, S. B. H. (1993). *Protein ladder sequencing. Science* **262,** 89–92.

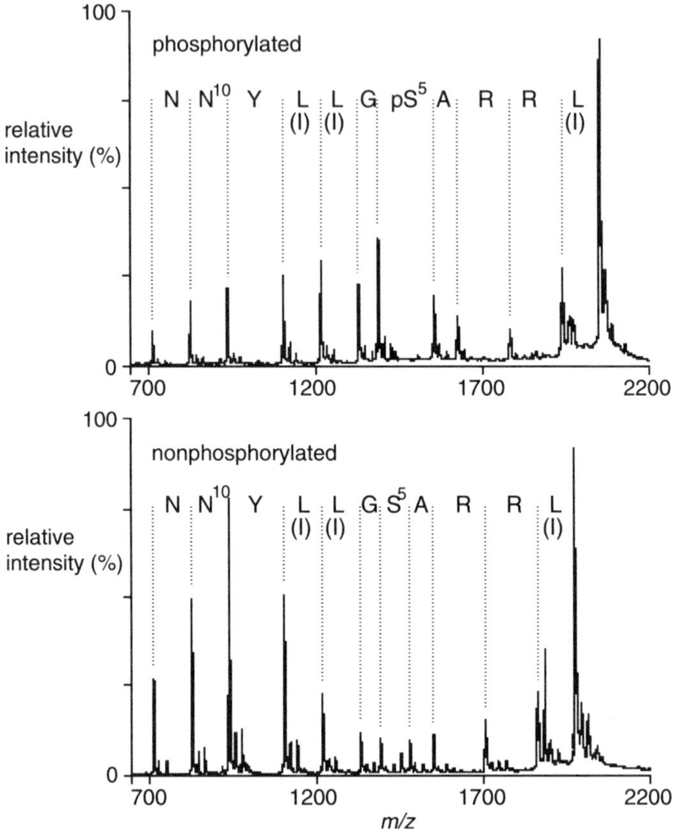

Figure 6.17 Protein ladder sequencing of a 16-residue synthetic peptide that is phosphorylated (top) and not phosphorylated (bottom). A partial amino acid sequence is noted above the mass spectra. (Adapted from Chait *et al.*, 1993).

Fitzgerald, M. C., Zhu, L., and Smith, L. M. (1993). *The analysis of mock DNA sequencing reactions using matrix-assisted laser desorption ionization mass spectrometry. Rapid Commun. Mass Spectrom.* **7,** 895–897.

Smith, L. M. (1993). *The future of DNA sequencing. Science* **262,** 530–532.

Mass Spectrometry and Viral Analysis

The preface of this book began with the statement that mass spectrometry is no longer confined to the analysis of small molecules. This last set of experiments briefly, but dramatically, illustrates this point.

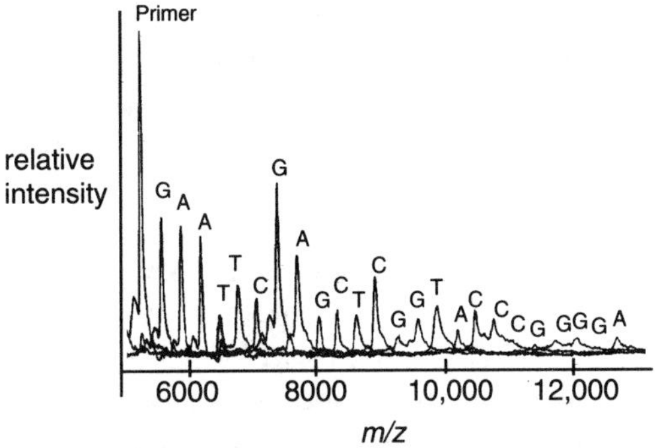

Figure 6.18 Negative ion MALDI mass spectra of synthetic oligonucleotides corresponding to mock A, C, G, and T sequencing reactions. The order of the peaks corresponds to the sequence. (Adapted from Fitzgerald *et al.*, 1993).

ESI mass spectrometry has generated a great deal of interest by allowing researchers to correlate gas phase biomolecular interactions with those in the condensed phase. Since the beginning of this century, mass spectrometry has progressed in its ability to analyze larger molecules and now, with the introduction of electrospray and MALDI, biological macromolecules have become readily amenable to mass analysis. In relation, virology has also experienced a progression to our current understanding of the pathogenesis of viruses and their assault on living organisms. Physical techniques such as X-ray crystallography and electron microscopy have greatly enriched our understanding of virus structure. The ability to accurately mass measure viral ions and other supramolecular complexes could significantly improve on the speed and accuracy of existing methods, with potential for distinguishing changes of 0.01% in the total mass of a complex.

Most macromolecular complexes studied by mass spectrometry are relatively small, on the order of 10^3–10^4 Da (Chapter 6), Monitoring Antibody Catalysis. Such studies rely on indirect evidence such as isotope exchange, charge state distribution, and changes in molecular weight to infer that native structure is retained in the gas phase. Lacking direct observation, the question remains whether native conformations are preserved during the ionization/vaporization process and the vacuum of the mass spectrometer. A second idea, reminiscent of the Manhattan Project where Calutron mass spectrometers were employed to separate uranium isotopes, is whether this technology can be used as a separation and collection device for biomolecules. Our work addresses these issues with the first direct observation of a supramolecular (4×10^7 Da) complex. Our ability to nondestructively generate, electrostatically focus, and mass select supramolecular complexes further demonstrates the feasibility of using mass spectrometry as a high-resolution separation/purification technique. This work suggests that there is potential for such preparative mass spectrometry.

Large ions can now be generated in the gas phase by electrospray and MALDI ion sources. However, the observation of these large ions is limited by the detection device. Detectors typically generate secondary electrons from colliding ions, which then cascade to generate more electrons, thereby resulting in an amplification on the order of $\sim 10^6$ electrons per striking ion. Large ions (>1,000,000 Da) do not produce significant secondary electron emission and therefore cannot be detected. In order to compensate for the limitations of the detector and test the range of the new ionization techniques, the detector was replaced and the ions were physically collected. The isolated specimen was directly examined by transmission electron microscopy. Figure 6.19 illustrates the basic design of the mass analysis instrumentation we have employed.

In addition to identifying the viruses following mass selection, the transmission electron micrographs also demonstrated preservation of the icosahedral rice yellow mottle virus (RYMV) capsid and the rod shape of the helical tobacco mosaic virus (TMV) capsid (Figure 6.20). The high molecular weight and characteristic rod-like structure of TMV provided a striking example of the the applicability of ESI–MS for viruses. Noteworthy is that the capsids of RYMV and TMV were stabilized by noncovalent interactions between protein subunits.

This work has important biochemical ramifications. Retention of virus ultrastructure provides direct evidence that the quaternary structure of noncovalent complexes can be conserved in the gas phase and throughout

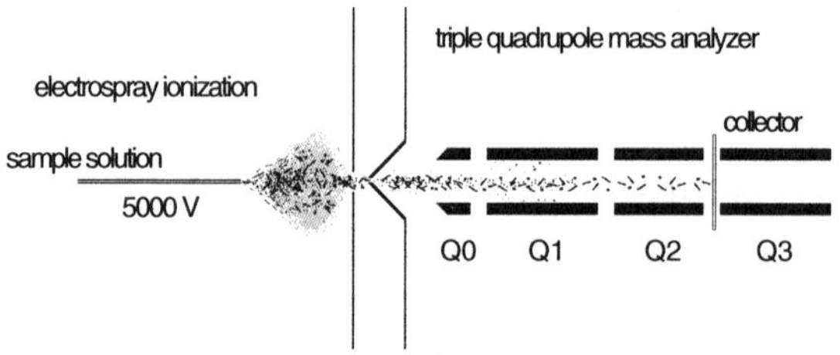

Figure 6.19 Electrospray ionization mass spectrometry with a collection device between quadrupoles Q2 and Q3.

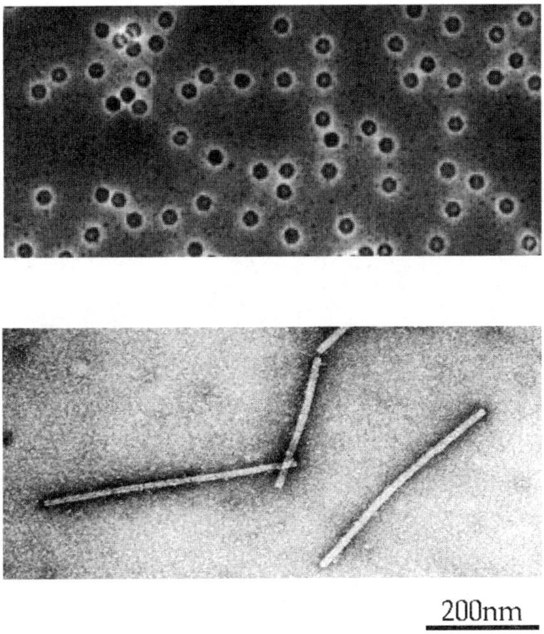

Figure 6.20 Rice yellow mottle virus (top) and tobacco mosaic virus (bottom) detected with electron microscopy after electroscopy ionization mass analysis.

the mass analysis process. Second, the ability to separate biomolecular complexes suggests that electrospray may be developed for preparative mass spectrometry as a way of performing high-resolution separation and purification.

Reference

Siuzdak, G., Bothner, B., Yeager, M., Brugidou, C., Fauquet, C. M., Hoey, K., and Chang, C.-M. (1995). Manuscript in preparation.

Terms and Definitions[1]

Adduct ion formation Adduct ion formation is commonly known as cationization or anionization. The noncovalent addition of an ion adduct involves the addition of a cation (H^+, Na^+, K^+, . . .) or an anion (Cl^-, I^-, SO_4^-, . . .) to a molecule. For example, [carbohydrate + Na]$^+$ is a common ion observed for carbohydrates in the positive ion analysis mode. It is also possible to observe neutral noncovalent adducts.

Cationization Cationization involves the noncovalent addition of a positively charged ion to a neutral molecule, resulting in a charged complex that can be observed by mass analysis. While protonation can be thought of as cationization, the term cationization is more commonly used for the addition of a cation adduct (alkali cations) other than a proton. Cationization is commonly used as a means of producing a stable molecular ion. For example, some molecules that are protonated fragment with little or no formation of the parent molecular ion. This is partially due to the strong binding between a proton and the molecule and thus the transfer of charge to the molecule. The charge transfer can destabilize the molecule and therefore promote fragmentation. Alkali cations can effectively "stick" to a molecule. Since the binding is not covalent in nature the charge remains localized and a more stable molecular ion is formed.

Collision-induced dissociation (CID) An ion/neutral process wherein the (fast) projectile ion is fragmented as a result of collision with a target neutral species. This process is known as collision-induced dissociation (CID) and/or collision-activated dissociation (CAD). CID (or CAD) is accomplished by conversion, during the collision, of part of the

[1] Some definitions were derived from P. Price (1991). Standard definitions of terms relating to mass spectrometry. A report from the committee on measurements and standards of the American Society for Mass Spectrometry. *J. Am. Soc. Mass Spectrom.* **2,** 336–348.

translational energy of the ion to internal energy in the ion. It is accomplished in tandem mass spectrometry by selecting an ion of interest with a mass filter/analyzer and introducing that ion into a collision cell. A collision gas (typically Ar) is also introduced into the collision cell, and the selected ion will collide with the neutral atoms within the cell, resulting in fragmentation. The fragments can then be analyzed to obtain a fragment ion (or daughter ion) spectrum.

Cosolvent A solvent is used to facilitate the solvation of an analyte into a FAB or MALDI matrix system. For example trifluoroacetic acid is used to facilitate solvation of some peptides into the FAB matrix, *m*-nitrobenzyl alcohol.

Derivatization Chemical alteration of a molecule to increase volatility and thermal stability by reducing intermolecular forces. Derivatization is a preparatory step for the thermal desorption process in electron ionization because it imparts volatility and thermal stability to molecules, making them more amenable to the thermal desorption. Derivatization will generally increase the molecular weight and reduce intermolecular forces by converting polar groups into funtionalities of lower polarity.

underivatized carbohydrate

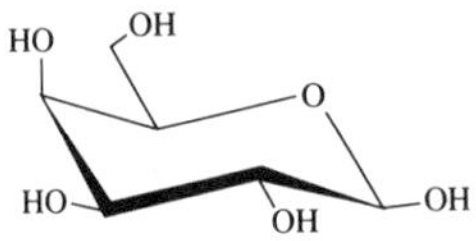

acetylated carbohydrate

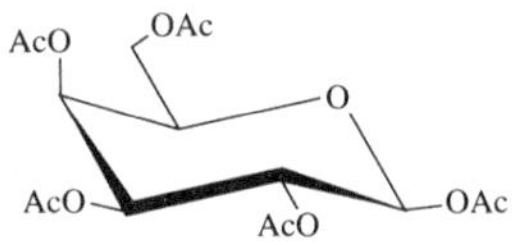

The reduction of intermolecular forces, such as hydrogen bonding, is accomplished by removing active hydrogen atoms. For example, acids, amides, or polyalcohols may be acetylated or trimethylsilylated. Electrostatic interactions of zwitterionic amino acids may also be quenched by acetylation and methylation.

Desorption ionization Desorption ionization is a general term used to encompass the various desorption ionization techniques (e.g., secondary ion MS, fast atom bombardment, plasma desorption, matrix-assisted laser desorption). Desorption ionization generates ions by desorbing them from a solid or liquid sample with a high-energy particle beam.

Electron ionization Ionization occurring through molecular interactions with electrons. The electrons, obtained from a heated filament, are accelerated through some voltage (typically 70 eV). The electrons transfer this kinetic energy to a molecule to produce ionization. Usually no more than an excess of 5–6 eV above the ionization potential (ionization potentials are usually 8–13 eV) is absorbed. The excess energy often results in significant fragmentation. The reaction

$$M + e^- \rightarrow M^{+\cdot} + 2e^-$$

is typical of the ionization that occurs in electron ionization. $M + e^- \rightarrow M^-$ also occurs in EI; however, it is usually less efficient.

Electrospray ionization (ESI) Electrospray ionization generates ions by spraying a solution (aqueous or organic solvent) through a charged inlet. Once the solvent is sprayed, its droplets rapidly desolvate through the addition of heat, a stream of gas (air), or both. As the solvent evaporates, ions in the highly charged droplets get ejected. The ions are then electrostatically directed into the mass analyzer. This ion source commonly produces multiply charged ions, making it easy to detect proteins with a quadrupole mass analyzer having a range of *m/z* 3000.

Exact mass This is also known as accurate mass measurement and means obtaining mass of an ion within an error of 5 to 30 ppm of the expected mass. Specific journal requirements are:

Journal of the American Chemical Society (JACS) Criteria are ±5 ppm for mass less than 1000 and unit mass for mass greater than 1000.

Journal of Organic Chemistry (JOC) Criteria are ±13 mmu for mass less than 500, ±16 millimass units for mass 500–1000, and unit mass for mass greater than 1000. For example:

Compound	Molecular formula and exact mass	Acceptable exact mass range (Da)	
		JACS	JOC
Buckyball	C_{60} = 720.0000 Da	719.9964 to 720.0036	719.9840 to 720.0160

Fragmentation Fragmentation is a process that occurs when enough energy is concentrated in a bond, causing the vibrating atoms to move a part beyond a bonding distance. In general, fragments resulting from weak bonds are prominent in the mass spectra.

Ionization and cationization Ionization of a neutral molecule occurs through the addition or removal of a charged species such as an electron or a proton. The following list describes how mass spectrometers achieve ionization.

Electron ejection The ejection of an electron to form a positive cation. Most ionization techniques will allow you to form this type of ion, but the most common is electron ionization.

Electron capture This is roughly 100 times less likely to occur in electron ionization; therefore it is 100 times less sensitive. The sensitivity is, however, very compound dependent. For example pentane will not easily accept an electron while hexafluorobenzene will.

Deprotonation A means of observing ions from acidic compounds. It is accomplished by removing a proton and monitoring them in the negative ion mode.

Protonation This is accomplished by the addition of a proton or protons to form a positively charged species. Protonation works well on many polar compounds. Proteins and peptides are well known for their ability to obtain a charge through protonation. Protonation may result in destabilization of certain molecules, for example carbohydrates; in these cases alkali cationization is useful.

Alkali cationization This is accomplished by the addition of an alkali cation. Many compounds are unstable to proton addition and require alkali cation complexation in order to observe stable molecular ions. Carbohydrates are well known to form stable alkali cationized molecular ions while their protonated molecular ions are often undetectable.

Ammonium cationization An alternative to alkali cationization.

Ion spray Pneumatically assisted electrospray is achieved by passing a stream of gas (air) over the droplets to facilitate desolvation.

Liquid secondary ion mass spectrometry (LSIMS) Essentially the same process as fast atom/ion bombardment (FAB), LSIMS originally derived from secondary ion mass spectrometry (SIMS). SIMS did not incorporate a liquid matrix.

Mass analyzers Mass analyzers separate ions according to their mass-to-charge ratio. A description of the more common analyzers is given below.

Electrostatic analyzer A velocity-focusing device for producing an electrostatic field perpendicular to the direction of ion travel (usually used in combination with a magnetic analyzer, below, for mass analysis). The effect is to bring to a common focus all ions that have been accelerated through a given voltage difference.

Ion trap analyzer A mass-resonance analyzer that produces a three-dimensional rotationally symmetric quadrupole field capable of storing ions at selected mass-to-charge ratios.

Magnetic sector analyzer A direction-focusing device that produces a magnetic field perpendicular to the direction of ion travel. The effect is to bring to a common focus all ions of a given mass-to-charge ratio.

Quadrupole mass analyzer A mass filter that creates a quadrupole field with a direct current component and a radio frequency component in such a manner as to allow scanning over a selected mass-to-charge range.

Time-of-flight analyzer A device that measures the flight time of ions over a fixed distance. The time is the same for ions that have been accelerated through the same voltage difference and have the same mass-to-charge ratio.

Mass measurement There are three different ways to calculate molecular weight. The way the mass is calculated and how that compares with the observed mass data depends largely on the accuracy and resolution of the mass spectrometer.

Average mass The mass of an ion calculated from a given empirical formula using the atomic weight (which is an average of the isotopes) for each element, such as C = 12.1115, H = 1.00797, O = 15.9994

Monoisotopic ion mass The mass of an ion calculated from a given empirical formula using the exact mass of the most abundant isotope of each element, such as C = 12.000000, H = 1.007825, O = 15.994915.

Nominal ion mass The mass of an ion calculated from a given empirical formula using the integer mass of the most abundant isotope of each element, such as C = 12, H = 1, O = 16.

Mass spectrometer An instrument in which ions are generated and analyzed according to their mass-to-charge ratio, and in which the number of ions is determined electrically.

Matrix-assisted laser desorption/ionization (MALDI) An ionization source that generates ions by desorbing them from a solid matrix material with a pulsed laser beam.

mDa MilliDaltons (0.001 Daltons).

mmu Millimass unit (0.001 Daltons).

m/z An abbreviation used to denote the dimensionless quantity formed by dividing the mass of an ion in Daltons by the number of charges carried by the ion. For example, for the ion $C_7H_7^{1+}$, $m/z = 91.0$. For the ion $C_7H_7^{2+}$, $m/z = 45.5$

Negative ion An atom, radical, molecule, or molecular moiety that has gained one or more electrons or lost a positive moiety, thus acquiring an electrically negative charge.

Nitrogen rule If a compound contains an even number of nitrogen atoms, its molecular ion will be an even mass number for compounds containing the elements C, H, N, and/or O and S. In this case a neutral compound having an even number of nitrogens will have an even number of hydrogens. Conversely, a compound having an odd number of nitrogens will typically have an odd number of hydrogens. Phosporus will behave as a nitrogen atom such that an even number of nitrogens and one phosphorus will result in an odd number of hydrogens. Halogens can be counted as hydrogens.

This rule can be useful when determining the molecular formula of your compound. For a neutral organic compound, elements C, H, N, and/or O and S, an odd number of nitrogens should result in an odd number of hydrogens and an even number of nitrogens should result in an even number of hydrogens.

Examples:

Compound	Formula	Monoisotopic mass (Da)
Methane	CH_4	14
Methyl amine	CH_5N	31
Methanol	CH_4O	32
Aniline	C_6H_7N	93
Thioglycerol	$C_3H_8O_2S_1$	108
Peptide (AARNDCCHIIP)	$C_{49}H_{81}N_{17}O_{15}S_2$	1211
Carbohydrate (Le^x)	$C_{23}H_{39}N_1O_{15}$	569
Oligonucleotide	$C_{23}H_{38}P_2N_4O_8$	560
Phospholipid	$C_{10}H_{23}P_1O_5$	254

Plasma desorption (californium fission fragment desorption) An ionization source that generates ions by desorbing them from a solid matrix material with high-energy fission fragments generated from radioactive californium.

Positive ion An atom, radical, molecule, or molecular moiety that has lost one or more electrons or gained one or more positive ions, thus acquiring an electrically positive charge.

ppm parts per million, a term often used in exact mass measurements to describe accuracy. For example, a 2.3-mDa error for a molecular ion with *m/z* 545.2034 is equivalent to a relative error of $0.0023 \div 545 = 0.0000042$ or 4.2×10^{-6} or 4.2 ppm.

actual = 545.2034
observed = 545.2011
error = 4.2 ppm

Resolution Resolving power, the ability of a mass spectrometer to separate two masses (M_1, M_2), is termed resolution (R). The most common definition of R is

$$R = M/\Delta M,$$

in which $\Delta M = M1 - M2$ and $M = M1 \approx M2$. Thus if a mass spectrometer can separate two masses (100,101) then $\Delta M = 1$, $M = 100$, and $R = 100$. For conventional accurate mass measurements, R needs to be as large as possible. An instrument with $R = 20{,}000$ can separate an ion at 100.0000 from a second mass at 100.0050 ($\Delta M = M/R = 100/20{,}000 = 0.005$).

To use the formula $R = M/\Delta M$, it is necessary to define at what stage the two peaks representing the two masses are actually separate. The height of the "valley" between the two peaks is used for this purpose, with 5, 10, and 50% valley definitions being in use. A 5%

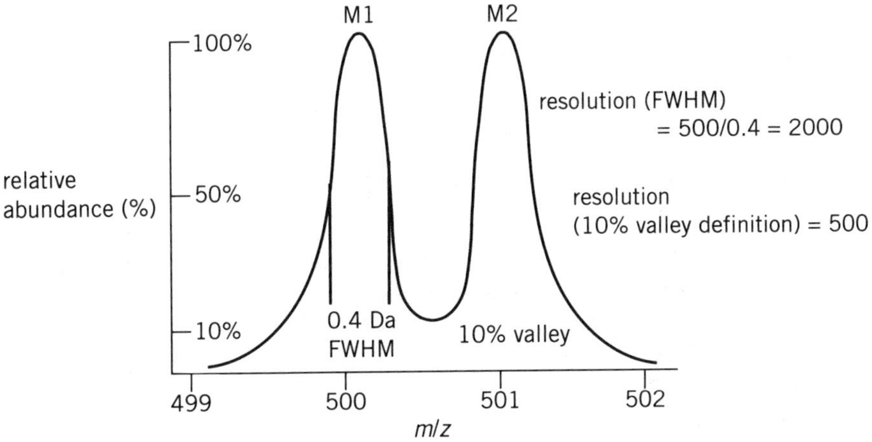

The application of two definitions of resolution.

valley definition is a much stricter criterion of separation (resolution) than is the 50% definition.

Resolution (10% valley definition) *M/ΔM* Let two peaks of equal height in a mass spectrum at masses m and $m - \Delta M$ be separated by a valley that at its lowest point is just 10% of the height of either peak. Then the resolution (10% valley definition) is $M/\Delta M$.

Resolution (peak width definition) *M/ΔM* For a single peak make up of singly charged ions of mass m in a mass spectrum, the resolution may be expressed as $M/\Delta M$, where ΔM is the width of the peak at a height that is a specified fraction of the maximum peak height. It is recommended that one of three values 50, 5, or 0.5% be used. For an isolated symmetrical peak, recorded with a system that is linear in range between the 5 and 10% levels of the peak, the 5% peak width definition is technically equivalent to the 10% valley definition. A common standard is the definition of resolution based upon ΔM being the full with of the peak at half its maximum height, sometimes abbreviated "FWHM."

Selected ion monitoring (SIM) This is also known as selected ion recording. Selecting a particular ion or ions to monitor has the benefit of sensitivity. Since the instrument is not spending time scanning for other ions, it can be dedicated to the ion selected. This can result in large increase in sensitivity; however, the disadvantage is a lack of information. It has been especially useful in biochemistry, medicine, and environmental science where a known compound or compounds required quantification from complex mixtures. SIM with derivatization has also been shown to be very useful with prostaglandins.

Signal-to-noise ratio (SNR) The signal to noise ratio, which is proportional to the square root of the number of samples, is averaged in scanning the signal. Thus

$$\text{SNR} \propto n^{1/2}.$$

So if a signal is averaged 100 times, the signal-to-noise ratio will increase by a factor of ten.

Tandem mass spectrometry Coupling two or more mass analyzers to perform repetitive mass spectrometry experiments. Typically used in conjunction with fragmentation experiments in which a particular ion is selected by the first analyzer, followed by collision-induced dissociation, and subsequent analysis of the fragment ions in the next analyzer.

Total ion current The total current generated by all the ions generated independent of m/z.

Appendix

TABLE A.1
Atomic Weights of the Elements Based on the Carbon 12 Standard

Symbol	Mass	Abundance	Symbol	Mass	Abundance	Symbol	Mass	Abundance
^{1}**H**	1.0078	99.985	^{48}Ca	47.9525	0.187	^{81}Br	80.9163	49.31
^{2}H	2.0141	0.015	^{45}**Sc**	44.9559	100.00	^{78}**Kr**	77.9204	0.35
^{3}**He**	3.0160	0.00014	^{46}**Ti**	45.9526	8.0	^{80}Kr	79.9164	2.25
^{4}He	4.0026	99.99986	^{47}Ti	46.9518	7.3	^{82}Kr	81.9135	11.60
^{6}**Li**	6.0151	7.5	^{48}Ti	47.9479	73.8	^{83}Kr	82.9141	11.50
^{7}Li	7.0160	92.5	^{49}Ti	48.9479	5.5	^{84}Kr	83.9115	57.00
^{9}**Be**	9.0122	100.00	^{50}Ti	49.9448	5.4	^{86}Kr	85.9106	17.30
^{10}**B**	10.0129	19.9	^{50}**V**	49.9472	0.250	^{85}**Rb**	84.9118	72.165
^{11}B	11.0093	80.1	^{51}V	50.9440	99.750	^{87}Rb	86.9092	27.835
^{12}**C**	12.0000	98.90	^{50}**Cr**	49.9460	4.35	^{84}**Sr**	83.9134	0.56
^{13}C	13.0034	1.10	^{52}Cr	51.9405	83.79	^{86}Sr	85.9093	9.86
^{14}**N**	14.0031	99.634	^{53}Cr	52.9407	9.50	^{87}Sr	86.9089	7.00
^{15}N	15.0001	0.366	^{54}Cr	53.9389	2.36	^{88}Sr	87.9056	82.58
^{16}**O**	15.9949	99.762	^{55}**Mn**	54.9380	100.00	^{89}**Y**	88.9059	100.00
^{17}O	16.9991	0.038	^{54}**Fe**	53.9396	5.80	^{90}**Zr**	89.9047	51.45
^{18}O	17.9992	0.200	^{56}Fe	55.9349	91.72	^{91}Zr	90.9056	11.27
^{19}**F**	18.9984	100.00	^{57}Fe	56.9354	2.20	^{92}Zr	91.9050	17.17
^{20}**Ne**	19.9924	90.51	^{58}Fe	57.9333	00.28	^{94}Zr	93.9063	17.33
^{21}Ne	20.9938	0.27	^{59}**Co**	58.9332	100.00	^{96}Zr	95.9083	02.78
^{22}Ne	21.9914	9.22	^{58}**Ni**	57.9353	68.27	^{93}**Nb**	92.9064	100.00
^{23}**Na**	22.9898	100.00	^{60}Ni	59.9308	26.10	^{92}**Mo**	91.9068	14.84
^{24}**Mg**	23.9850	78.99	^{61}Ni	60.9311	1.13	^{94}Mo	93.9051	9.25
^{25}Mg	24.9858	10.00	^{62}Ni	61.9283	3.59	^{95}Mo	94.9058	15.92
^{26}Mg	25.9826	11.01	^{64}Ni	63.9280	0.91	^{96}Mo	95.9047	16.68
^{27}**Al**	26.9815	100.00	^{63}**Cu**	62.9296	69.17	^{97}Mo	96.9060	9.55
^{28}**Si**	27.9769	92.23	^{65}Cu	64.9278	30.83	^{98}Mo	97.9054	24.13
^{29}Si	28.9765	4.67	^{64}**Zn**	63.9291	48.60	^{100}Mo	99.9075	9.63
^{30}Si	29.9738	3.10	^{66}Zn	65.9260	27.90	^{96}**Ru**	95.9076	5.52
^{31}**P**	30.9738	100.00	^{67}Zn	66.9271	4.10	^{98}Ru	97.9053	1.88
^{32}**S**	31.9721	95.02	^{68}Zn	67.9248	18.80	^{99}Ru	98.9059	12.70
^{33}S	32.9715	0.75	^{70}Zn	69.9253	0.60	^{100}Ru	99.9042	12.60
^{34}S	33.9679	4.21	^{69}**Ga**	68.9256	60.10	^{101}Ru	100.9056	17.00
^{36}S	35.9671	0.02	^{71}Ga	70.9247	39.90	^{102}Ru	101.9043	31.60
^{35}**Cl**	34.9689	75.77	^{70}**Ge**	69.9242	20.50	^{104}Ru	103.9054	18.70
^{37}Cl	36.9660	24.23	^{72}Ge	71.9221	27.40	^{103}**Rh**	102.9055	100.00
^{36}**Ar**	35.9675	0.337	^{73}Ge	72.9235	7.80	^{102}**Pd**	101.9056	1.02
^{38}Ar	37.9627	0.063	^{74}Ge	73.9212	36.50	^{104}Pd	103.9040	11.14
^{40}Ar	39.9264	99.600	^{76}Ge	75.9214	7.80	^{105}Pd	104.9051	22.33
^{39}**K**	38.9626	93.2581	^{75}As	74.9216	100.00	^{106}Pd	105.9035	27.33
^{40}K	39.9637	0.0117	^{74}**Se**	73.9225	0.90	^{108}Pd	107.9039	26.46
^{41}K	40.9618	6.7302	^{76}Se	75.9192	9.00	^{110}Pd	109.9052	11.72
^{40}**Ca**	39.9626	96.941	^{77}Se	76.9199	7.60	^{107}**Ag**	106.9051	51.839
^{42}Ca	41.9586	0.647	^{78}Se	77.9173	23.50	^{109}Ag	108.9048	48.161
^{43}Ca	42.9588	0.135	^{80}Se	79.9165	49.60	^{106}**Cd**	105.9065	1.25
^{44}Ca	43.9555	2.086	^{82}Se	81.9167	9.40	^{108}Cd	107.9042	0.89
^{46}Ca	45.9537	0.004	^{79}Br	78.9183	50.69	^{110}Cd	109.9030	12.49

TABLE A.1 (*Continued*)

Symbol	Mass	Abundance	Symbol	Mass	Abundance	Symbol	Mass	Abundance
^{111}Cd	110.9042	12.80	^{142}Ce	141.9092	11.08	^{174}**Hf**	173.9401	0.16
^{112}Cd	111.9028	24.13	^{141}**Pr**	140.9077	100.00	^{176}Hf	175.9414	5.20
^{113}Cd	112.9044	12.22	^{142}**Nd**	141.9077	27.13	^{177}Hf	176.9432	18.60
^{114}Cd	113.9034	28.73	^{143}Nd	142.9098	12.18	^{178}Hf	177.9437	27.10
^{116}Cd	115.9048	7.49	^{144}Nd	143.9101	23.80	^{179}Hf	178.9458	13.74
^{113}**In**	112.9041	4.30	^{145}Nd	144.9126	8.30	^{180}Hf	179.9466	35.20
^{115}In	114.9039	95.70	^{146}Nd	145.9131	17.19	^{180}**Ta**	179.9475	0.012
^{112}**Sn**	111.9048	1.00	^{148}Nd	147.9169	5.76	^{181}Ta	180.9467	99.988
^{114}Sn	113.9028	0.70	^{150}Nd	149.9209	5.64	^{180}**W**	179.9467	0.13
^{115}Sn	114.9033	0.40	^{144}**Sm**	143.9120	3.10	^{182}W	181.9482	26.30
^{116}Sn	115.9017	14.70	^{147}Sm	146.9149	15.00	^{183}W	182.9502	14.30
^{117}Sn	116.9030	7.70	^{148}Sm	147.9148	11.30	^{184}W	183.9510	30.67
^{118}Sn	117.9016	24.30	^{149}Sm	148.9172	13.80	^{186}W	185.9544	23.60
^{119}Sn	118.9033	08.60	^{150}Sm	149.9173	7.40	^{185}**Re**	184.9530	37.40
^{120}Sn	119.9022	32.40	^{152}Sm	151.9197	26.70	^{187}Re	186.9558	62.60
^{122}Sn	121.9034	4.60	^{154}Sm	153.9209	22.70	^{184}**Os**	183.9525	0.02
^{124}Sn	123.9053	5.60	^{151}**Eu**	150.9199	47.80	^{186}Os	185.9539	1.58
^{121}**Sb**	120.9038	57.30	^{153}Eu	152.9212	52.20	^{187}Os	186.9558	1.60
^{123}Sb	122.9042	42.70	^{152}**Gd**	151.9198	0.20	^{188}Os	187.9559	13.30
^{120}**Te**	119.9040	0.096	^{154}Gd	153.9209	2.18	^{189}Os	188.9582	16.10
^{122}Te	121.9031	2.60	^{155}Gd	154.9226	14.80	^{190}Os	189.9585	26.40
^{123}Te	122.9043	0.908	^{156}Gd	155.9221	20.47	^{192}Os	191.9615	41.00
^{124}Te	123.9028	4.816	^{157}Gd	156.9240	15.65	^{191}**Ir**	190.9606	37.30
^{125}Te	124.9044	7.14	^{158}Gd	157.9241	24.84	^{193}Ir	192.9629	62.70
^{126}Te	125.9033	18.95	^{160}Gd	159.9271	21.86	^{190}**Pt**	189.9599	0.01
^{128}Te	127.9045	31.69	^{159}**Tb**	158.9254	100.00	^{192}Pt	191.9610	0.79
^{130}Te	129.9062	33.80	^{156}**Dy**	155.9243	0.06	^{194}Pt	193.9627	32.90
^{127}**I**	126.9045	100.00	^{158}Dy	157.9244	0.10	^{195}Pt	194.9648	32.80
^{124}**Xe**	123.9061	0.10	^{160}Dy	159.9252	2.34	^{196}Pt	195.9649	25.30
^{126}Xe	125.9043	0.09	^{161}Dy	160.9269	18.90	^{198}Pt	197.9679	7.20
^{128}Xe	127.9035	1.91	^{162}Dy	161.9268	25.50	^{197}**Au**	196.9666	100.00
^{129}Xe	128.9048	26.40	^{163}Dy	162.9287	24.90	^{196}**Hg**	195.9658	0.15
^{130}Xe	129.9035	4.10	^{164}Dy	163.9292	28.20	^{198}Hg	197.9668	10.10
^{131}Xe	130.9051	21.20	^{165}**Ho**	164.9303	100.00	^{199}Hg	198.9683	17.00
^{132}Xe	131.9041	26.90	^{162}**Er**	161.9288	0.14	^{200}Hg	199.9683	23.10
^{134}Xe	133.9054	10.40	^{164}Er	163.9292	1.61	^{201}Hg	200.9703	13.20
^{136}Xe	135.9072	8.90	^{166}Er	165.9303	33.60	^{202}Hg	201.9706	29.65
^{133}**Cs**	132.9054	100.00	^{167}Er	166.9321	22.95	^{204}Hg	203.9735	6.80
^{130}**Ba**	129.9063	0.106	^{168}Er	167.9324	26.80	^{203}**Ti**	202.9723	29.524
^{132}Ba	131.9050	0.101	^{170}Er	169.9355	14.90	^{205}Ti	204.9744	70.476
^{134}Ba	133.9045	2.417	^{169}**Tm**	168.9342	100.00	^{204}**Pb**	203.9730	1.40
^{135}Ba	134.9057	6.592	^{168}**Yb**	167.9339	0.13	^{206}Pb	205.9745	24.10
^{136}Ba	135.9046	7.854	^{170}Yb	169.9348	3.05	^{207}Pb	206.9759	22.10
^{137}Ba	136.9058	11.23	^{171}Yb	170.9363	14.30	^{208}Pb	207.9766	52.40
^{138}Ba	137.9052	71.70	^{172}Yb	171.9364	21.90	^{209}**Bi**	208.9804	100.00
^{138}**La**	137.9071	0.09	^{173}Yb	172.9382	16.12	^{232}**Th**	232.0381	100.00
^{139}La	138.9064	99.91	^{174}Yb	173.9389	31.80	^{234}**U**	234.0409	0.0055
^{136}**Ce**	135.9071	0.19	^{176}Yb	175.9426	12.70	^{235}U	235.0439	0.7200
^{138}Ce	137.9060	0.25	^{175}**Lu**	174.9408	97.40	^{238}U	238.0508	99.2745
^{140}Ce	139.9054	88.48	^{176}Lu	175.9427	2.60			

TABLE A.2
Amino Acids and Their Masses Organized in Alphabetical Order

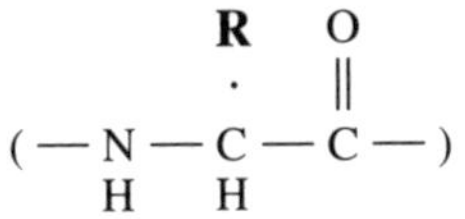

Amino acid	Letter code		Mass	R
		Alphabetical order		
Alanine	Ala	A	71	$-CH_3$
Arginine	Arg	R	156	$-CH_2(CH_2)_2NH-C=NHNH_2$
Asparagine	Asn	N	114	$-CH_2CONH_2$
Aspartic acid	Asp	D	115	$-CH_2COOH$
Cysteine	Cys	C	103	$-CH_2SH$
Glutamic acid	Glu	E	129	$-CH_2CH_2COOH$
Glutamine	Gln	Q	128	$-CH_2CH_2CONH_2$
Glycine	Gly	G	57	$-H$
Histidine	His	H	137	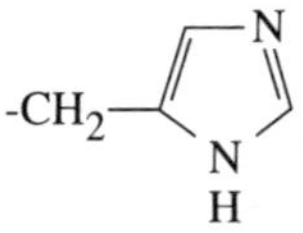
Isoleucine	Ile	I	113	$-CH(CH_3)CH_2CH_3$
Leucine	Leu	L	113	$-CH_2CH(CH_3)_2$
Lysine	Lys	K	128	$-CH_2(CH_2)_3NH_2$
Methionine	Met	M	131	$-CH_2CH_2SCH_3$
Phenylalanine	Phe	F	147	$-CH_2$-phenyl
Proline	Pro	P	97	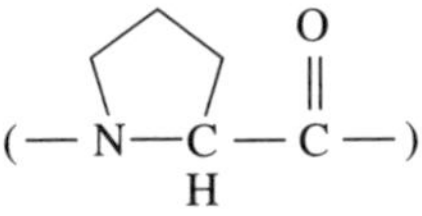
Serine	Ser	S	87	$-CH_2OH$
Threonine	Thr	T	101	$-CH(OH)CH_3$
Tryptophan	Trp	W	186	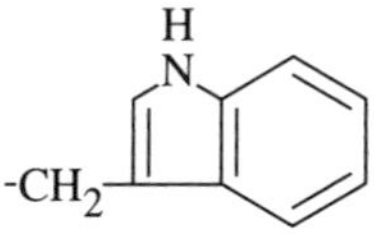
Tyrosine	Tyr	Y	163	$-CH$-*para*-phenol
Valine	Val	V	99	$-CH(CH_3)_2$

TABLE A.3
Amino Acids and Their Masses Organized According to Molecular Weight

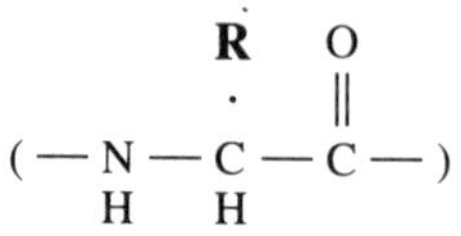

Amino acid	Letter code		Mass	**R**
			By mass	
Glycine	Gly	G	57	$-H$
Alanine	Ala	A	71	$-CH_3$
Serine	Ser	S	87	$-CH_2OH$
Proline	Pro	P	97	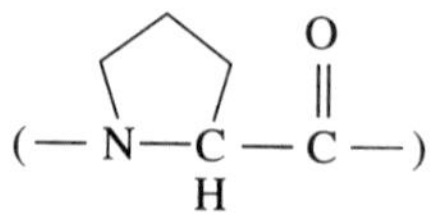
Valine	Val	V	99	$-CH(CH_3)_2$
Threonine	Thr	T	101	$-CH(OH)CH_3$
Cysteine	Cys	C	103	$-CH_2SH$
Isoleucine	Ile	I	113	$-CH(CH_3)CH_2CH_3$
Leucine	Leu	L	113	$-CH_2CH(CH_3)_2$
Asparagine	Asn	N	114	$-CH_2CONH_2$
Aspartic acid	Asp	D	115	$-CH_2COOH$
Glutamine	Gln	Q	128	$-CH_2CH_2CONH_2$
Lysine	Lys	K	128	$-CH_2(CH_2)_3NH_2$
Glutamic acid	Glu	E	129	$-CH_2CH_2COOH$
Methionine	Met	M	131	$-CH_2CH_2SCH_3$
Histidine	His	H	137	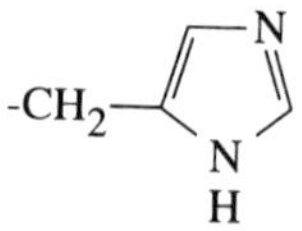
Phenylalanine	Phe	F	147	$-CH_2$-phenyl
Arginine	Arg	R	156	$-CH_2(CH_2)_2NH-C=NHNH_2$
Tyrosine	Tyr	Y	163	$-CH_2$-*para*-phenol
Tryptophan	Trp	W	186	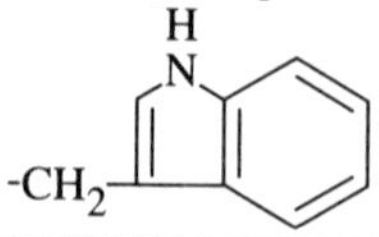

TABLE A.4
FAB Matrix Ions Observed with the NBA, NBA/NaI, and NBA/CsI Matrix Systems

NBA	*m/z*	NBA/NaI	*m/z*	NBA/CsI	*m/z*
$NBA + H^+ - OH$	137	$NBA + H^+ - OH$	137	Cs^+	133
$NBA + H^+$	154	$NBA + H^+$	154	$NBA + H^+$	154
$NBA_2 + H^+ - H_2O$	289	Na_2I^+	173	$NBA + Cs^+$	286
$NBA_2 + H^+$	307	$NBA + Na^+$	176	$NO_2 + Cs_2^+$	312
$NBA_3 + H^+ - H_2O$	443	$NBA_2 + H^+ - H_2O$	289	Cs_2I^+	393
$NBA_3 + H^+$	460	$NBA_2 + H^+$	307	$NBA + Cs_2^+$	419
$NBA_4 + H^+$	613	$Na_3I_2^+$	326	$NBA_2 + Cs^+$	439
$NBA_5 + H^+$	766	$NBA_2 + Na^+$	329	$NBA + Cs_2I^+$	546
$NBA_6 + H^+$	919	$NBA_3 + H^+ - H_2O$	443	$NBA_2 + Cs_2^+$	572
$NBA_7 + H^+$	1072	$NBA_3 + H^+$	460	$NBA_3 + Cs^+$	592
		$Na_4I_3^+$	479	$Cs_3I_2^+$	653
		$NBA_3 + Na^+$	482	$NBA + Cs_3I^+$	679
		$NBA_4 + H^+$	613	$NBA_2 + Cs_2I^+$	699
		$NBA_4 + Na^+$	635	$NBA_3 + Cs_2^+$	725
				$NBA + Cs_3I_2^+$	806
				$NBA_2 + Cs_3I^+$	832
				$Cs_4I_3^+$	912
				$NBA + Cs_4I_2^+$	938
				$NBA_2 + Cs_4I_2^+$	1091
				$Cs_5I_4^+$	1172
				$NBA + Cs_5I_3^+$	1198
				$Cs_6I_5^+$	1432

TABLE A.5
General Comparison of Ionization Techniques: Electron Ionization (EI), Fast Atom/Ion Bombardment (FAB), Matrix-Assisted Laser Desorption/Ionization (MALDI), Electrospray Ionization (ESI)

	Mass limit (practical)	Advantages	Disadvantages	Suitable compounds (See Chapters 3–6)
EI	~1,000 (500)	Wide availability Utility of fragmentation Sensitivity (high fmole) High resolution	Limited to low mass range Can be too much fragmentation Thermal desorption	Small relatively nonpolar molecules
FAB	24,000 (7000)	Wide availability Easily adapted to high-resolution instruments MCI capable of multiple charging	Salt interference LC/MS flow rates limited to ~2 μl/min Matrix required for LC/MS Low sensitivity Matrix interference	Peptides Carbohydrates Nucleotides Small polar molecules
MALDI	>300,000 (150,000)	Tolerance of mM salts Tolerance of mixtures Femtomole sensitivity Promise as a tool for sequence analysis	Low resolution Matrix interference Low accuracy	Peptides Proteins Glycoproteins Carbohydrates Nucleotides Oligonucleotides Phosphoproteins Small polar molecules
ESI	~200,000 (70,000)	Multiple charging LC/MS capability Softest ionization Femtomole sensitivity	Multiple charging can be confusing with mixtures Salt sensitivity	Peptides Proteins Glycoproteins Carbohydrates Nucleotides Oligonucleotides Phosphoproteins Small polar molecules

TABLE A.6
MALDI or ESI?

	ESI (single quad)	ESI (triple quad or ion trap)	MALDI (linear)	MALDI (reflectron)
Practical mass limit	70,000 Da	70,000 Da	300,000 Da	300,000 Da
Typical resolution	2000	2000	150–500	~2000 for cpds <6000 Da ~400 for cpds >6000 Da
Accuracy	0.01%	0.01%	0.01% to 0.1%	0.01% to 0.1%
Sensitivity	Subpicomole	Subpicomole	Subpicomole	Subpicomole
MS/MS capabilities	Limited	Full	None	Limited
Direct LC compatibility	Fully	Fully	None	None
Contaminant tolerance	Somewhat	Somewhat	Very	Very
Sample preparation	Direct from salt-free solution	Direct from salt-free solution	Add sample to *appropriate* matrix	Add sample to *appropriate* matrix
Ease of use	Easy	Easy	Easy	Easy
Cost	$140,000–180,000	$150,000–350,000	$75,000–100,000	$150,000–250,000
Peptide modifications	Good	Ideal	Poor	OK
Purity and homogeneity determination	Qualitative	Qualitative	Qualitative	Qualitative
Noncovalent interactions	Possible	Possible	Not possible	Not possible
Quantitation	With internal standard	With internal standard	With internal standard	With internal standard

Note. MALDI and ESI have emerged as very useful tools in biological research. The question "Which instrument should I buy?" often arises. In many cases the techniques are complementary; however, often a choice must be made. This table may help determine what technique best suits your needs.

Index